Marlis Hofer

Statistical downscaling of atmospheric variables

Marlis Hofer

Statistical downscaling of atmospheric variables

for data-sparse, glaciated mountain sites

Südwestdeutscher Verlag für Hochschulschriften

Impressum/Imprint (nur für Deutschland/only for Germany)
Bibliografische Information der Deutschen Nationalbibliothek: Die Deutsche Nationalbibliothek verzeichnet diese Publikation in der Deutschen Nationalbibliografie; detaillierte bibliografische Daten sind im Internet über http://dnb.d-nb.de abrufbar.
Alle in diesem Buch genannten Marken und Produktnamen unterliegen warenzeichen-, marken- oder patentrechtlichem Schutz bzw. sind Warenzeichen oder eingetragene Warenzeichen der jeweiligen Inhaber. Die Wiedergabe von Marken, Produktnamen, Gebrauchsnamen, Handelsnamen, Warenbezeichnungen u.s.w. in diesem Werk berechtigt auch ohne besondere Kennzeichnung nicht zu der Annahme, dass solche Namen im Sinne der Warenzeichen- und Markenschutzgesetzgebung als frei zu betrachten wären und daher von jedermann benutzt werden dürften.

Coverbild: www.ingimage.com

Verlag: Südwestdeutscher Verlag für Hochschulschriften GmbH & Co. KG
Heinrich-Böcking-Str. 6-8, 66121 Saarbrücken, Deutschland
Telefon +49 681 37 20 271-1, Telefax +49 681 37 20 271-0
Email: info@svh-verlag.de

Approved by: Innsbruck, LFU, Diss., 2012

Herstellung in Deutschland (siehe letzte Seite)
ISBN: 978-3-8381-2974-7

Imprint (only for USA, GB)
Bibliographic information published by the Deutsche Nationalbibliothek: The Deutsche Nationalbibliothek lists this publication in the Deutsche Nationalbibliografie; detailed bibliographic data are available in the Internet at http://dnb.d-nb.de.
Any brand names and product names mentioned in this book are subject to trademark, brand or patent protection and are trademarks or registered trademarks of their respective holders. The use of brand names, product names, common names, trade names, product descriptions etc. even without a particular marking in this works is in no way to be construed to mean that such names may be regarded as unrestricted in respect of trademark and brand protection legislation and could thus be used by anyone.

Cover image: www.ingimage.com

Publisher: Südwestdeutscher Verlag für Hochschulschriften GmbH & Co. KG
Heinrich-Böcking-Str. 6-8, 66121 Saarbrücken, Germany
Phone +49 681 37 20 271-1, Fax +49 681 37 20 271-0
Email: info@svh-verlag.de

Printed in the U.S.A.
Printed in the U.K. by (see last page)
ISBN: 978-3-8381-2974-7

Copyright © 2012 by the author and Südwestdeutscher Verlag für Hochschulschriften GmbH & Co. KG and licensors
All rights reserved. Saarbrücken 2012

TABLE OF CONTENTS

ACKNOWLEDGEMENTS 4

ABSTRACT 5

ZUSAMMENFASSUNG 6

CHAPTER

    I. Overview 7

        1.1 Introduction 7
            1.1.1 Outline of the thesis 7
            1.1.2 Why focus on the Cordillera Blanca, a glaciated mountain range in the tropics? 8
            1.1.3 The applied statistical procedures within the framework of atmospheric downscaling techniques 10
            1.1.4 Objectives 12
        1.2 Abstracts of the papers 14
            1.2.1 Paper 1: Empirical-statistical downscaling of reanalysis data to high-resolution air temperature and specific humidity above a glacier surface (Cordillera Blanca, Peru) 14
            1.2.2 Paper 2: Skill assessment of NCEP/NCAR reanalysis data for daily air temperature on a glaciated mountain range (Peru) 15
            1.2.3 Paper 3: Comparing the skill of different reanalyses as predictors for daily air temperature on a glaciated mountain (Peru) 15
        1.3 Benefits of the thesis 16
        1.4 Outlook 17

    II. Paper 1: Empirical-statistical downscaling of reanalysis data to high-resolution air temperature and specific humidity above a glacier surface (Cordillera Blanca, Peru) 19

2.1	Introduction	19
2.2	Study site and observations	21
2.3	The empirical-statistical downscaling (ESD) model	22
	2.3.1 What is ESD and why do we need it?	22
	2.3.2 Choice of predictor fields	23
	2.3.3 Transfer functions for individual times of day	27
	2.3.4 Empirical orthogonal function (EOF) analysis and predictor pre-processing	28
	2.3.5 PC screening based on cross-validation	28
	2.3.6 Model calibration and double cross-validation	30
2.4	Results and discussion	30
	2.4.1 Predictor analysis	30
	2.4.2 Discussion of the month/hour-models	32
	2.4.3 Cross-validation of the month/hour-models	34
	2.4.4 Performance of the ESD model against simpler reference models	35
	2.4.5 ESD model forecasts 1960 to 2008	40
2.5	Conclusions	43

III. Paper 2: Skill assessment of NCEP/NCAR reanalysis data for daily air temperature on a glaciated mountain range (Peru) — 44

3.1	Introduction	44
3.2	Study site and observations: the predictands	47
3.3	ESD model architecture	49
	3.3.1 Accounting for seasonal periodicity	49
	3.3.2 Predictor selection	50
	3.3.3 Downscaling process: linear model calibration and cross-validation	52
	3.3.4 Example	55
3.4	ESD model application	55
	3.4.1 Downscaling parameters	55
	3.4.2 Skill assessment	58
	3.4.3 Results for different time scales	60
	3.4.4 Towards automated predictor selection	62
	3.4.5 Accounting for the effects of diurnal periodicity	63
3.5	Summary and conclusions	64

IV. Paper 3: Comparing the skill of different reanalyses as predictors for daily air temperature on a glaciated mountain (Peru) — 67

4.1	Introduction	67
4.2	Data	70

4.3	The method		71
4.4	Results		74
	4.4.1	Optimum scale analysis	75
	4.4.2	Comparing the skill of different reanalyses	79
4.5	Conclusions		81
	4.5.1	Results confined to the case study	81
	4.5.2	General recommendations	81

BIBLIOGRAPHY 83

LIST OF FIGURES 94

ACKNOWLEDGEMENTS

This work was funded by the "Verein zur Förderung der wissenschaftlichen Tätigkeit von Südtirolern an der Landesuniversität Innsbruck", and by the Austrian Science Foundation FWF (P22106-N21).

Grateful acknowledgement is made to P. Wagnon (IRD) for kindly providing the data from Artesonraju glacier (paper 1).

NCEP reanalysis data are freely provided by the National Oceanic and Atmospheric Administration (NOAA), Boulder, Colorado, USA, from their Web site at http://www.esrl.noaa.gov/psd/ (papers 1, 2 and 3). ERA-interim have been obtained from the European Centre for Medium-range Weather Forecasts (ECMWF). Data-sets have been used from the JRA-25 long-term reanalysis cooperative research project carried out by the Japan Meteorological Agency (JMA) and the Central Research Institute of Electric Power Industry (CRIEPI). MERRA are disseminated by the Global Modeling and Assimilation Office (GMAO) and the Goddard Earth Sciences Data and Information Services Center (GES DISC) (paper 3).

Besonderer Dank gilt meinem Betreuer Georg Kaser, für seine Unterstützung und sein Vertrauen, seine Motivation und das Eröffnen von immer neuen Perspektiven. Diese Dissertation wäre nicht möglich gewesen ohne die fruchtreiche Zusammenarbeit mit Ben Marzeion und Thomas Mölg. Vielen Dank für die zahlreichen Diskussionen und hilfreichen Ratschläge! An dieser Stelle möchte ich auch die anderen Mitglieder dieser wunderbaren Arbeitsgruppe erwähnen: Stephan Galos, Martin Großhauser, Alex Jarosch, Irmgard Juen, Lindsey Nicholson, Rainer Prinz, und Michael Winkler. Jeder einzelne hat vieles zum Gelingen der Dissertation beigetragen.

Ganz großen Dank möchte ich meinen Eltern Lorenz und Burgl aussprechen, weil sie mein Studium immer befürwortet haben.

Vielen Dank an meinen lieben Mattia, für seine zahlreichen Tipps, seine permanente Unterstützung, und seine Geduld.

ABSTRACT

How much information about local-scale weather can be gained from large-scale atmospheric models? The thesis investigates downscaling methods for temporally high-resolution (sub-daily to daily), and normally distributed target variables. The performed procedures, based on model output statistics, are particularly appropriate when only short-term observations (i.e., few-years) are available. Specific attention is given to objective predictor selection and evaluation. An assessment of skill is included that primarily focuses on intra-seasonal (day-to-day) and inter-annual (year-to-year) variability.

The elaborated methods are applied to a case study, with air temperature and humidity records from automatic weather stations in the Cordillera Blanca (Peru) representing the target variables, and reanalysis data from different institutions the predictors. The goal is to extend information available from the existing short-term observations into the past, in order to provide a better understanding about high-resolution atmospheric variability in the Cordillera Blanca. The downscaling model skill shows important seasonal variations, with high skill in the synoptically active, rainy season (January to March), and low skill in the dry season (June to August; where generally smaller intra-seasonal and inter-annual variations occur). A comparison of reanalyses from several institutions yields notable differences in the performances of the different products. E.g., reanalyses based on 4-dimensional variational analysis show considerably higher skill than those based on 3-dimensional variational analysis, and differences in skill are evident especially during dry season months. Results of the optimum scale analysis reveal that horizontal grid point averaging effectively enhances the downscaling model, as it reduces numerical noise related to single grid points in the large-scale models.

In principle, the downscaling techniques presented here are portable to different sites or variables, as long as a few years of observational data are available for training. Yet we recommend their use primarily for skill assessment of large-scale predictors, and the subsequent application of results in impact studies only if high skill is demonstrated.

ZUSAMMENFASSUNG

Wieviel Information über das lokale Wettergeschehen steckt in grobskaligen Atmosphärenmodellen? Ziel dieser Dissertation ist die Ausarbeitung von Downscalingmethoden für zeitlich hochaufgelöste, normalverteilte Atmosphärenvariablen (Stunden- oder Tageswerte). Die Downscalingmethoden beruhen auf statistischen Verfahren, die speziell für kurze Beobachtungszeitreihen (wenige Jahre) geeignet sind. Ein Modellvalidierungsverfahren in Bezug auf Variabilität von Tag zu Tag (intra-seasonal) und Jahr zu Jahr (inter-annual) wird vorgestellt, mit besonderem Augenmerk auf die objektive Auswahl und Evaluierung der Prädiktoren (unabhängige Variablen aus dem Atmoshärenmodell).

Die statistischen Methoden werden anhand einer Fallstudie demonstriert. Zielvariablen sind Lufttemperatur- und Feuchtemessungen von automatischen Wetterstationen in der Cordillera Blanca (Peru), und Prädiktoren Reanalysedaten verschiedener Institutionen. Ziel der Fallstudie ist es, auf nur kurzzeitig verfügbaren Messzeitreihen aufbauend mehr über das längerfristige Wettergeschehen in der Cordillera Blanca zu lernen. Die Ergebnisse zeigen eine deutlich größere Vorhersagbarkeit der Zielvariablen durch die grobskaligen Reanalysedaten in der Regenzeit (Jänner bis März), als in der Trockenzeit (Juni bis August). Ein Vergleich von verschiedenen Reanalysedaten zeigt, dass Produkte basierend auf vierdimensionaler Variationsanalyse wesentlich bessere Prädiktoren sind (besonders in der Trockenzeit), als jene basierend auf dreidimensionaler Variationsanalyse. Weiters wird eine Analyse der räumlichen Skalen durchgeführt, über welche gemittelt die Prädiktoren die besten Ergebnisse erzielen. Diese zeigt, dass durch horizontale Mittelung numerische Fehler in den Gitterpunktdaten verringert, und somit Modellergebnisse wesentlich verbessert werden können.

Prinzipiell können die in dieser Dissertation ausgearbeiteten Downscalingmethoden beliebig auf andere Untersuchungsgebiete und Variablen übertragen werden, wenn kurzzeitige Messreihen für die Kalibrierung des statistischen Modells zur Verfügung stehen. Wir empfehlen allerdings die Verwendung der Methoden in erster Linie für Evaluierung, und die konkrete Anwendung der Modelldaten nur, wenn eine entsprechend gute Bewertung vorliegt.

CHAPTER I

Overview

1.1 Introduction

1.1.1 Outline of the thesis

An overview section, and a collection of three research papers, constitute my thesis submitted in partial fulfillment of the requirements for the degree of Doctor rerum naturalium at the Faculty for Geo- and Atmospheric Sciences (Leopold Franzens University of Innsbruck, Austria).

The overview (chapter I) provides (1) an introduction to the most important aspects and goals of the thesis, (2) a summary of the papers, and (3) conclusions and outlook. The second part (chapters II, III, and IV) consists of three papers submitted to, or published by, international peer review journals. These papers are:

- **Paper 1:** Hofer, M., T. Mölg, B. Marzeion, and G. Kaser (2010), Empirical-statistical downscaling of reanalysis data to high-resolution air temperature and specific humidity above a glacier surface (Cordillera Blanca, Peru), *Journal of Geophysical Research*, 115 (D12120), 15.

- **Paper 2:** Hofer, M., B. Marzeion, and T. Mölg (2011), Skill assessment of NCEP/NCAR reanalysis data for daily air temperature on a glaciated mountain range (Peru), *Journal of Climate*, under review since 6/2011.

- **Paper 3:** Hofer, M., B. Marzeion, and T. Mölg (2011), Comparing the skill of different reanalyses as predictors for daily air temperature on a glaciated mountain (Peru), *Climate Dynamics*, under review since 7/2011.

Paper 1 was honored with the Leopold Franzens University of Innsbruck Best Student Paper Award in 2011.

1.1.2 Why focus on the Cordillera Blanca, a glaciated mountain range in the tropics?

The Cordillera Blanca (figure 1.1) is Peru's most extensively glaciated mountain range, and it harbors 25% of all tropical glaciers with respect to surface area (e.g., *Kaser and Osmaston*, 2002). Over the 20th century, in accordance with global trends, Cordillera Blanca glaciers have shrunk substantially (e.g., *Ames*, 1998; *Georges*, 2004; *Silverio and Jaquet*, 2005; *Racoviteanu et al.*, 2008). Yet these glaciers represent important hydrological components in the highly populated Rio Santa Valley, as they contribute significantly to balancing the high runoff seasonality (e.g., *Mark and Seltzer*, 2003; *Kaser et al.*, 2003; *Juen et al.*, 2007; *Baraer et al.*, 2009; *Kaser et al.*, 2010). More specifically, the glaciers store a surplus of precipitation in the wet season and release melt water in the dry season, when almost no precipitation occurs (*Niedertscheider*, 1990). This way perennial availability of water resources has been maintained over time, upon which societies have become increasingly dependent (e.g., *Juen*, 2006; *Mark et al.*, 2010; *Carey*, 2010; *Bury et al.*, 2011).

Juen et al. (2007) present an estimate of future mass balance for glaciers in the Cordillera Blanca, and the hydrologic implications. They conclude that under climatic changes as projected by IPCC scenarios, the glaciers will continue to decrease, and consequently the seasonality in the runoff will intensify considerably. In a case study based on a few-years measurements in the Yanamarey watershed (located in the Southern Cordillera Blanca), *Bury et al.* (2011) draw similar conclusions. Thus Cordillera Blanca inhabitants can be regarded as particularly vulnerable to water stress problems related to climatic change (e.g., *Juen et al.*, 2007; *Mark et al.*, 2010; *Bury et al.*, 2011; *Chevallier et al.*, 2011). In addition to their importance as water resources, Cordillera Blanca glaciers have heavily impacted the socioeconomic development in the region by causing some of the most disastrous catastrophes recorded in history, in terms of outburst floods and avalanches (*Carey*, 2005, 2010).

The glacier-climate relationship in the Cordillera Blanca has been the focus of numerous studies (e.g., *Kaser et al.*, 1990; *Kaser and Georges*, 1997; *Mark and Seltzer*, 2003; *Kaser et al.*, 2003; *Georges*, 2005; *Kaser et al.*, 2005; *Juen*, 2006; *Vuille et al.*, 2008). All these studies, however, have been restricted to the assessment of monthly mean atmospheric data, since higher-resolution measurements are not available on a longer term. Since 1999, though, temporally high-resolution measurements of air temperature, humidity, pressure, wind, and radiation are recorded at an automatic weather station (AWS) network at, and around, glaciers in the Cordillera Blanca (the

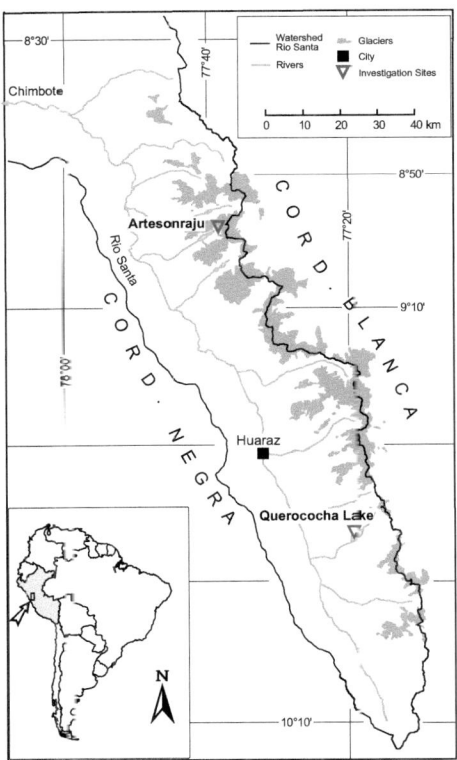

Figure 1.1: Map of the Cordillera Blanca with main glaciers and the Rio Santa water shed (the indicated positions of the glacier Artesonraju and the lake Querococha are referred to in chapter II).

sites of the stations are indicated in figure 3.1; for a more detailed description please refer to the work of *Juen*, 2006). These few years of data provide a valuable basis for process-oriented studies of glacier mass balance variations (e.g., *Mölg et al.*, 2009) that help understanding the intricate tropical glacier-climate interactions (e.g., *Kaser*, 1995; *Kaser et al.*, 1996; *Kaser and Osmaston*, 2002; *Juen et al.*, 2011). However, maintaining the AWSs to provide continuous and reliable measurements has been a logistical challenge and the time series include substantial data gaps (*Juen*, 2006; *Hofer et al.*, 2010). This, in combination with the short time period of measurements, limits the utility of the data sets in providing information about longer-term, past atmospheric variations.

1.1.3 The applied statistical procedures within the framework of atmospheric downscaling techniques

"Even if global climate models in the future are run at high resolution there will remain the need to 'downscale' the results from such models to individual sites or localities for impact studies" (*DOE*, 1996)

Ongoing developments in atmospheric modeling have made available choices of long-term, temporally high-resolution atmospheric data sets on a global scale. These data, however, are still restricted to coarse spatial scales, such that their immediate application to study regional and local climate is not recommended. Especially over complex topography, such as mountains, the large-scale atmospheric models miss significant processes that characterize local weather.

Methods to infer local-scale information, as required e.g. in impact assessments, from global atmospheric models are known as downscaling methods. Statistical-type downscaling (*Klein et al.*, 1959) has existed since the early development of computer-based, numerical atmospheric modeling. These methods adjust atmospheric model output by empirically determining relationships between a set of model variables (the predictors), and observed variables (the predictands). About a decade later, *Hill* (1968) proposed a completely alternative downscaling approach that involves the solution of numerical equations on a smaller-scale, limited-area grid nested into the larger-scale model, to obtain the desired grid resolution for the region of interest within acceptable computational expenses. This type of limited-area modeling is also called dynamical downscaling. Since then, significant progress has been made both in statistical and in dynamical downscaling, and a variety of different models and approaches in both classes has emerged.

Dynamical downscaling is based on the basic atmospheric equations of motion, and mass and energy conservation, similar to the global models, and therefore has the potential of providing physically consistent, regional-scale information along multiple climate variables. Problems regarding the implementation of lateral boundary conditions in limited area modeling are well documented (e.g., *Davies*, 1976; *Warner et al.*, 1997), and some studies use nudging or relaxation techniques to minimize distortion of the large scales (e.g, *Von Storch et al.*, 2000b; *Miguez-Macho et al.*, 2004). To enable quantitative estimates of many sources of uncertainties in regional climate projections, ensemble simulations (*Christensen et al.*, 2002) are being used more and more. However, a major drawback of dynamical models is their high computational expense that still limits applications in terms of simulation period and resolution (e.g., *Mölg and Kaser*, 2011).

Statistical downscaling is computationally inexpensive. Three main classes exist (*Wilby et al.*, 2002), namely (1) weather typing, (2) stochastic weather generators, and (3) transfer functions. All these methods rely upon the unverifiable stationarity assumption, i.e. the assumption that the statistical model is valid under altered climatic conditions (note that this is a concern also in dynamical downscaling, regarding the parameterizations in the limited area models, e.g., *McFarlane*, 2011). Regarding the predictor data, statistical downscaling methods are also categorized in two fundamentally different approaches, namely perfect prognosis, or Perfect Prog (*Klein et al.*, 1959), and model output statistics (MOS, *Glahn and Lowry*, 1972). Further details about this distinction are given in chapter III, figure 3.2.

Weather typing involves classifying weather patterns based on large-scale model output, and grouping observed, local-scale meteorological data conditional on these weather patterns. The observed data distributions for each pattern are than associated with the same patterns in periods with no observations. Analog models are an example of weather classifications where predictands are chosen by matching previous weather states. Analog models were originally proposed by *Lorenz* (1969), but not renewed until by *Zorita et al.* (1995) and *Martin et al.* (1996), with the upcoming availability of long predictor time series, in terms of reanalysis data (e.g., *Kalnay et al.*, 1996), which are essential to make these models feasible. A further weather classification approach is the use of hidden Markov models (e.g., *Hughes and Guttorp*, 1994).

Weather generators (e.g., *Richardson*, 1981; *Wilks and Wilby*, 1999; *Qian et al.*, 2005) stochastically produce synthetic daily or sub-daily time series of consistent sets of weather variables that conform observed statistics. Time series of the same variables are then generated for unobserved periods by perturbing parameters of the weather generator according to changes as projected by the driving large-scale model. Most weather generators first simulate rainfall occurrence and then coherently sets of other climate variables, conditional on rainfall. A criticism of the perhaps best known approach, the Richardson-type weather generator (*Richardson*, 1981), is its failure to adequately describe the length of dry and wet series which is important in many applications (e.g., agricultural impacts). To overcome this problem, an alternative, the "serial approach", has evolved, which first models the sequence of dry and wet series and then the other weather variables. Weather generators can be used whenever long or spatially gridded time series of daily weather are required. However, weather generators need a considerable amount of observations (e.g., 30 years of daily data) for adjusting its parameters. Another problem is that the spatially gridded data from weather generators are usually site-independent (i.e., they do not include any information about spatial correlations of weather variables). E.g., a drought might appear less severe, because the dry sequences simulated by a weather generator occur only at one site and not simultaneously at different sites, even if droughts are commonly a widespread phenomenon.

Transfer-function downscaling relies on empirical predictor-predictand relationships based on linear or non-linear, single or multiple regression (including also logistic regression and artificial neural networks). A variety of methodologies exist, often pragmatically implemented to serve specific application needs and not readily transferable to distinct tasks. Approaches largely differ in terms of choice of the statistical transfer function, or of the applied predictor variables. A particular problem related to regression-based downscaling approaches is that they, as a matter of fact, underestimate the observed variability (e.g., *Von Storch*, 1999).

All the above mentioned statistical methods include the major drawback that generally large observational data sets are necessary for the model training.

1.1.4 Objectives

One major goal of this thesis is the development of statistical downscaling methods that are easily transferrable to distinct applications. In the first instance, however, the study has been motivated by the requirement of providing results, i.e. high-resolution atmospheric data in the Cordillera Blanca (Peru), and thus the application of techniques at a specified site. This requirement has determined the basic architecture of statistical procedures presented in this thesis.

Major challenges to the statistical downscaling techniques in this thesis, as posed by the present case study, can be summarized as

1. the fact that high-resolution measurements for the model training and validation are available in the Cordillera Blanca only on a short term (i.e., few years), which constrains or even inhibits the use of many existing downscaling techniques (see above),

2. the location of the study site within very complex orography that largely differs from what global climate models are able to represent today, including extremely sharp topographic gradients (with 27 peaks reaching above 6000 m a.s.l. in an area of only about 30 km longitudinal and 180 km latitudinal extend), as well as small-scale land cover variability (e.g., glaciers next to non glaciated terrain), and

3. the situation of the study site in the tropics, where atmospheric variability inferred from large-scale data sets is more uncertain (e.g., Karl et al., 2006; Trenberth et al., 2007), given the scarcity of observational data to be incorporated in production and validation of these data, and that relatively few research has been conducted to increase the general process-understanding, or to quantify these uncertainties (compared to northern hemisphere mid-latitudes).

Based on the introductory remarks above, the main objectives of the thesis can be viewed from two sides:

(a) the objective of general method development in statistical downscaling, and

(b) the objective of gaining knowledge about high-resolution, past atmospheric variability in the Cordillera Blanca.

Thus the goals of the thesis are (with the numbering referring to the three papers that constitute the main part of this thesis, and **(a)** and **(b)** to the above classification):

1. to find or design a downscaling technique appropriate for short-term and high-resolution meteorological data **(a)**, and its application to extend available meteorological time series from a high-altitude, glaciated site in the Cordillera Blanca to the past **(b)**,

2. the development of an objective predictor selection and/or comparison tool for statistical downscaling applications (again, suitable in cases when only short-term, high-resolution observations are available) **(a)**, to be demonstrated by determining the most appropriate predictors for Cordillera Blanca air temperature predictands **(b)**, and

3. the application of this method to identify the most skillful large-scale data set, out of available products, for daily air temperature in the Cordillera Blanca. **(b)**.

1.2 Abstracts of the papers

1.2.1 Paper 1: Empirical-statistical downscaling of reanalysis data to high-resolution air temperature and specific humidity above a glacier surface (Cordillera Blanca, Peru)

Recently initiated observation networks in the Cordillera Blanca (Peru) provide temporally high-resolution, yet short-term atmospheric data. The aim of this study is to extend the existing time series into the past. We present an empirical-statistical downscaling (ESD) model that links 6-hourly NCEP/NCAR reanalysis data to air temperature and specific humidity, measured at the tropical glacier Artesonraju (Northern Cordillera Blanca). The ESD modeling procedure includes combined empirical orthogonal function and multiple regression analyses, and a double cross-validation scheme for model evaluation. Apart from the selection of predictor fields, the modeling procedure is automated and does not include subjective choices. We assess the ESD model sensitivity to the predictor choice using both single- and mixed-field predictors. Statistical transfer functions are derived individually for different months and times of day. The forecast skill largely depends on month and time of day, ranging from 0 to 0.8. The mixed-field predictors perform better than the single-field predictors. The ESD model shows added value, at all time scales, against simpler reference models (e.g., the direct use of reanalysis grid point values). The ESD model forecast 1960 to 2008 clearly reflects inter-annual variability related to the El Niño/Southern Oscillation, but is sensitive to the chosen predictor type.

1.2.2 Paper 2: Skill assessment of NCEP/NCAR reanalysis data for daily air temperature on a glaciated mountain range (Peru)

How much information about local-scale atmospheric variations can be gained from large-scale climate models? The study presents an empirical-statistical downscaling (ESD) MOS-technique (model output statistics) that is designed for objective predictor selection in the case of short-term, daily time series of normally distributed predictand variables. We estimate the ESD model skill using a cross-validation scheme that accounts for persistence in the time series. Seasonal periodicity in the time series is avoided by using separate models for each calendar month.

In this study, the ESD procedure is applied to NCEP reanalysis data predictors for daily air temperature predictands, measured at high-altitude sites on a glaciated mountain range (Peru). The results show high seasonality in both ESD parameters and model skill, emphasizing the importance of using separate models for each month. NCEP predictors show high skill in the wet season, but low skill in the dry season. We conclude that in the wet season the stronger synoptic forcing leads to higher predictability based on large-scale predictors, whereas during the dry season, small-scale radiation processes dominate the local variability. A non data-based (a priori) selected predictor (air temperature in the pressure levels of the study site) clearly shows higher skill than other potential predictors, such as air temperature at the NCEP model surface or sea level pressure. We find that spatial averaging of large-scale predictors increases the ESD model skill as it reduces errors related to single grid point data.

1.2.3 Paper 3: Comparing the skill of different reanalyses as predictors for daily air temperature on a glaciated mountain (Peru)

It is well known from previous research that significant differences exist amongst reanalysis products from different institutions. Here, we compare the skill of NNRP (the National Centers for Environmental Prediction/National Center for Atmospheric Research reanalyses), ERA-int (the European Centre of Medium-range Weather Forecasts Interim), JCDAS (the Japanese Meteorological Agency Climate Data Assimilation System reanalyses), and MERRA (the Modern Era Retrospective-Analysis for research and Applications by the National Aeronautics and Space Administration) as predictors for daily air temperature on a high-altitude glaciated mountain site in Peru.

First, linear regression models between the large-scale predictors and the target variable are calibrated individually for different calendar months. Then, a skill estimation method especially suited for short-term, high-resolution time series is employed, which relies on cross-validation under consideration of persistence in the time series. The most important findings are: 1) differences in reanalysis skill appear especially during dry and intermediate seasons; 2) reanalysis systems including 4-dimensional variational analysis (i.e., ERA-int and MERRA) show considerably higher skill than those based on 3-dimensional variational analysis (NNRP and JCDAS); 3) the optimum horizontal scales largely vary between the different reanalyses, and horizontal grid resolutions of the reanalyses are weak indicators of this optimum scale; and 4) using a too large domain has less negative impact on the reanalysis performance, than using a too small domain.

1.3 Benefits of the thesis

Similar to the objectives classification defined above, also the outcomes of this thesis can be distinguished in **(a)** general method development, and **(b)** results specific to the present case study.

The contributions with respect to **(a)** can be summarized:

In paper 1, a novel methodology of transfer-function downscaling is presented, which is easily applied and especially suited for high-resolution (daily to sub-daily) and short-term (few-years) time series, because it accounts for periodicity and autocorrelation in the cross-validation. A speciality of the model training/evaluation procedure is that first a relatively large predictor pool can be input to the model, which then automatically selects the most important modes thereof (based on principal component analysis and a double cross-validation scheme), thereby avoiding subjective choices. However, the problem of predictor selection (a general issue in statistical downscaling, see chapter II), is not fully solved, since also the large predictor input first needs to be constrained. Furthermore, the results indicate that the downscaling outputs depend upon the predictor choice.

In paper 2, the above mentioned problems are met by proposing an objective procedure for predictor selections and model output evaluations, designed for normally distributed, short-term and high-resolution atmospheric time series.

Outcomes of paper 3 include recommendations regarding predictor selection in statistical downscaling. E.g., the assessment of optimum scales indicates that spatial averaging of grid points can significantly increase the skill of large-scale model output. It is important to note that the minimum scales (i.e. distance between two neighboring model grid points) are not good indicators for this optimum scale. Further results suggest that the more modern reanalysis data, which include 4-dimensional variational analysis, show considerably higher skill than reanalysis data based on 3-dimensional variational analysis.

With regard to (b), the following conclusions can be made.

In paper 1, the presented downscaling model is applied to extend Cordillera Blanca sub-daily air temperature and humidity time series to the past. The model shows considerable added value over simpler reference models and is capable of reflecting inter-annual variations related to the El Niño Southern Oscillation. For sub-daily timescales, the model skill varies depending on calendar month and hour of day, and it must be noted that even if high skill is found in some months and hours, the underestimation of variability in other months and hours limits the applicability of results in impact studies.

In paper 2, the presented predictor selection tool is applied to Cordillera Blanca daily air temperature time series, giving more insight to problems encountered in paper 1. While the large-scale data show even surprisingly high skill during wet-season months the skill reduces almost to zero during the dry season. This can be explained by large-scale atmospheric activity more importantly affecting local-scale variations in the region during the wet season, while during the dry season, the generally smaller variability is dominated by local-scale noise.

Results in paper 3 show that, from all available global reanalyses, ERA-interim (see chapter IV for a reference) are the most skillful predictors for daily air temperature in the Cordillera Blanca. Improvements compared to the the other reanalysis data are evident especially in the dry and intermediate seasons, where ERA-interim show considerably higher skill.

1.4 Outlook

In my thesis, statistical downscaling methods with focus on predictor selection, evaluation and processing are presented. These methods are applied to extend high-resolution atmospheric time series in the Cordillera Blanca to unobserved periods. Results are encouraging, showing that even very coarse-scale model output (e.g.,

NCEP reanalyses, see chapter II for a reference) reflects small-scale atmospheric variations up to sub-daily time scales. However, results also show a considerable portion of unexplained variability, which complicates the application of the downscaling results. Beyond this case study, I recommend the usage of the downscaling procedures primarily for skill assessment of large-scale data sets, and the application of downscaling outputs only when considerable skill can be proved based on the here presented evaluation techniques. In order to further increase process-understanding about large- and local-scale atmospheric variations in the Cordillera Blanca, and their inter-relationship, high-resolution limited area modeling is considered useful (e.g., Mölg et al., 2011). The most appropriate lateral boundary data for the limited area model experiments can be chosen, e.g., using the predictor selection tools presented here.

CHAPTER II

Paper 1: Empirical-statistical downscaling of reanalysis data to high-resolution air temperature and specific humidity above a glacier surface (Cordillera Blanca, Peru)

Marlis Hofer, Thomas Mölg, Ben Marzeion, and Georg Kaser

Faculty of Geo- and Atmospheric Sciences, University of Innsbruck (Austria)

(published 2010 in *Journal of Geophysical Research*, 115/D12120, 15)

2.1 Introduction

Mountain glaciers are widely recognized as sensitive indicators of climatic change and variability. Glaciers in the tropics are of particular interest, because they respond faster to climatic changes than glaciers in the mid- or high latitudes (*Kaser*, 1995). Located at high altitudes, tropical glaciers moreover provide important climate information about the tropical free troposphere. Knowledge about climate change in this zone is incomplete, because the observational data are sparse and controversial (e.g., *Kaser*, 1995; *Karl et al.*, 2006; *Trenberth et al.*, 2007). More than 99% of all tropical glaciers (in terms of surface area) are located in the South American Andes, and roughly 70% in Peru (e.g., *Kaser and Osmaston*, 2002). Peru's most extensively glacier covered mountain range is the Cordillera Blanca that harbors 25% of the tropical glaciers. Figure 1.1 shows a map of the Cordillera Blanca with the main glaciers and the Rio Santa water shed. Similar to the global trends, substantial glacier shrinkage in the Cordillera Blanca over the 20th century has been documented (e.g., *Ames*, 1998; *Georges*, 2004; *Silverio and Jaquet*, 2005). Modeling studies predict that glaciers in the Cordillera Blanca will continue to shrink in the future, with drastic consequences for the runoff (e.g., *Juen*, 2006; *Juen et al.*, 2007). The runoff from glaciers is of great socioeconomic concern in this region, because it provides an important service as seasonal water regulator (e.g., *Kaser et al.*, 2003). During the dry season, almost all fresh water used by agriculture, industry and households in the Rio Santa valley originates from glaciers (e.g., *Mark and Seltzer*, 2003). To estimate potential impacts of climate change on glacier systems in the Cordillera Blanca is thus of primary importance for water resource management in this extensively populated and developing region (e.g., *Juen*, 2006).

The formulation of the energy balance at the glacier surface represents a deterministic approach to quantifying atmospheric controls of glacier mass loss (e.g. *Kuhn*, 1989). Surface energy balance studies conducted on tropical glaciers (e.g. *Favier et al.*, 2004; *Hastenrath*, 1978; *Mölg and Hardy*, 2004; *Wagnon et al.*, 1999) reveal that, apart from air temperature, tropical glaciers are particularly sensitive to moisture (e.g., *Kaser et al.*, 2005). Surface energy balance investigations in the Cordillera Blanca are generally limited to short periods, because they require a variety of on-site observations at high temporal resolution for associated modeling. Measurements on the remote tropical high glaciers, however, are logistically difficult to carry out and have therefore remained rare. In 1999 the ITGG (Innsbruck Tropical Glaciology Group, Austria) and the IRD (Institut de Recherche pour le Développement, France) initiated an observation network on glaciers in the Cordillera Blanca by installing automatic weather stations (AWSs).

The goal of this study is to extend high-resolution, but short-term meteorological time series on tropical glaciers to the past. Therefore, we present an empirical-statistical downscaling (ESD) scheme for translating coarse-scale reanalysis data to the local meteorological conditions. Our approach is particular in the context of ESD for mainly two reasons. First, the data series for calibration cover only a very short period (approximately two years, see further section 2.2). Second, unlike most ESD studies (e.g., *Benestad et al.*, 2008), we aim to forecast variables at a very high temporal resolution (i.e., 6-hourly values). Our target variables are two of the key drivers in the surface energy balance of tropical glaciers: air temperature and specific humidity, here above a tropical glacier surface in a very complex orographic setting. These two variables govern turbulent heat exchange to a large extent (e.g., *Mölg and Hardy*, 2004), and also influence radiative energy (e.g., *Mölg et al.*, 2009).

We want to explore the possibilities and limitations of ESD with regard to reaching the aims stated above. Study site and observational data are introduced in section 2.2. We describe the employed ESD modeling procedure (including the predictor selection) in section 2.3 The results of the ESD model are presented and evaluated in section 2.4. Finally, we summarize our conclusions in section 2.5.

2.2 Study site and observations

The investigation area in this study is the glacier Artesonraju, located at 8.9°S and 77.4°W in the Paron Valley (Northern Cordillera Blanca, cf. figure 1.1). The glacier covers an area of about 5.7 km^2 and reaches from 6025 m a.s.l. down to approximately 4750 m a.s.l. (*Juen*, 2006). In 2004, the ITGG and the IRD installed four AWSs at and around the glacier. The maintenance of these stations has been difficult due to logistical problems. To date, approximately two-year records are available from each station and it is not decided yet if they can be maintained in the future. In this study we mainly focus on data from one AWS at approximately 4850 m a.s.l. operated by the IRD (hereafter simply referred to as AWS), because it is the only of the four AWSs that is located on the glacier and that is equipped with air temperature and humidity sensors (additionally to wind speed/direction, incoming/outgoing short- and longwave radiation). The measurement period at AWS extends from March 2004 to May 2006 (hereafter referred to as calibration period). Due to several instruments failures the data series are not continuous in time. Within the calibration period, data are missing from November 2004 to February 2005 (approximately 15% of the data; periods of missing data are marked in figures 2.7 and 2.8 presented later). We refer to the work by *Juen* (2006) for a detailed description of the study site and measurement set-up.

In the entire Cordillera Blanca only one long-term air temperature record exists, from a station operated by Electroperu at lake Querococha (3980 m a.s.l., 77.3°W 9.72°S, about 100 km distant from the glacier Artesonraju, see figure 1.1). The observations at Querococha are available from 1965 to 1994 (only monthly means), and are presented in section 2.4.5, where annual means of the ESD model forecast 1960 to 2008 are discussed.

2.3 The empirical-statistical downscaling (ESD) model

2.3.1 What is ESD and why do we need it?

General circulation models (GCMs) are powerful tools to simulate past, present, and future climate and weather, but their spatially coarse resolution does not realistically resolve local and regional atmospheric states (e.g., *Grotch and MacCracken*, 1991). This is true also for reanalysis data (introduced later), which are closely related to GCM output. Hence techniques have emerged to downscale GCM output to higher spatial resolutions as required for climate change impact studies. Figure 2.1 depicts the role of downscaling to transfer GCM output (and reanalysis data, respectively) to higher spatial scales. The development of downscaling strategies is an important focus in regional climate research (e.g., *Christensen et al.*, 2007). Two different approaches exist: empirical-statistical downscaling (ESD), and dynamical downscaling. Dynamical downscaling is comprehensive limited area modeling with initial and lateral boundary conditions provided by the GCM. Yet the limited area models are computationally expensive: for long-term simulations their spatial resolutions are restricted to >10 km. The added value of regional climate models over GCMs is still under debate (e.g., *Prömmel et al.*, 2010) and usually the regional climate model output needs further downscaling steps before it can be applied in impact studies (e.g., *Früh et al.*, 2006; *Salzmann*, 2006).

ESD, the technique applied in this study, is a statistical model that transfers large-scale model output to the local scale. One advantage of ESD against dynamical downscaling is its much lower computational demand. However, it can be applied only for sites where observations are available for model calibration. A number of ESD techniques exist for widely differing applications and some of them focus on mountain or glacier environments (e.g., *Matulla*, 2005; *Radic and Hock*, 2006; *Reichert and Bengtsson*, 2002; *Salzmann*, 2006). However, most ESD studies concentrate on variables in the monthly time resolution (*Benestad et al.*, 2008). We aim for high temporal resolutions, because a further scope of this study is to provide input data for process-based glacier mass balance models required at sub-daily time steps (e.g., *Mölg et al.*, 2009). Thus, in terms of temporal resolution, our ESD approach is novel in climate research. In the context of numerical weather prediction similar techniques have been explored for decades (perfect prog technique; *Klein et al.*, 1959). With regard to downscaling at a sub-daily time scale in complex and glacier-covered mountain environments, however, this study represents a completely new approach.

In its simplest form, ESD can be expressed as the random or deterministic function

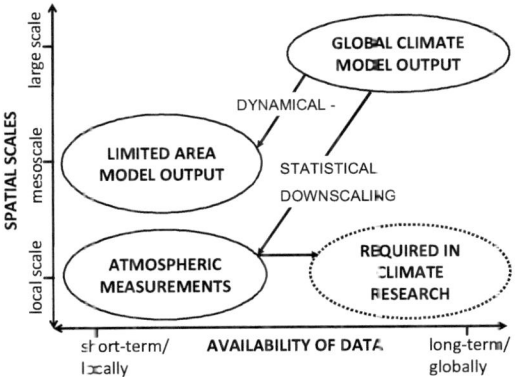

Figure 2.1: The role of downscaling techniques to transfer atmospheric data with high availability in time and space (abscissa), but low spatial resolution (ordinate), to the higher spatial resolutions required in climate research. In this figure we use the more general term "global climate model output", but it is true also for reanalysis data.

f, such that

$$Y = f(X) + \epsilon(t) \tag{2.1}$$

where Y is the predictand (the local scale target variable), X the large scale predictor, and ϵ a random error (or noise) (e.g. Von Storch et al., 2000a). In this study, the predictands Y are 6-hourly variations of air temperature and specific humidity above the surface of the glacier Artesonraju (measured at AWS). As predictors, X, we use reanalysis data.

2.3.2 Choice of predictor fields

Reanalysis data are the state-of-the-art archive of past meteorological variables spanning the entire atmosphere. They are produced by data assimilation, with a modeling system kept frozen over the simulation period and using quality controlled observations. Reanalysis data are available e.g. from the National Centers for Environmental Prediction/National Center for Atmospheric Research (NCEP/NCAR) (Kalnay et al., 1996), the European Centre for Medium-Range Weather Forecasts (ECMWF) (Uppala et al. 2005), or the Japan Meteorological Agency (JMA) (Kazutoshi et al., 2007). In this study we use the NCEP/NCAR reanalysis data, because they have shown skill in several studies focusing on the South American Andes (e.g. *Garreaud et al.*, 2003). NCEP/NCAR reanalysis data are available in sub-daily time resolution (6-hourly) for the period from 1948 to present, with a spatial resolution of approximately 210 km (T62), and 28 'sigma' levels in the vertical.

Only a few studies have systematically assessed the skill of different predictors in terms of variable types (e.g., *Cavazos and Hewitson*, 2005) or downscaling domain (e.g., *Brinkmann*, 2002), and there is no consensus on the most appropriate choice (*Fowler et al.*, 2007). ESD is based on the following three assumptions for suitable predictors (e.g., *Benestad et al.*, 2008). The predictors must (i) have a physical relationship to the predictand, (ii) be reliably represented by the reanalysis data or GCM, and (iii) reflect climate change. In this study, criterion (i) is specified further as a linear physical relationship, because we apply linear regression techniques. Criterion (ii) refers to the reliability of variables in the NCEP/NCAR reanalysis data. According to the availability of observations in the data assimilation, the NCEP/NCAR reanalysis variables are classified as type A, B, and C (*Kalnay et al.*, 1996). Type A variables (such as upper air temperature, wind and geopotential height) are most reliable, because they include measurements. Type C variables (e.g. surface fluxes, heating rates, or precipitation) are determined completely by the model and therefore associated with larger uncertainty. Moisture and surface variables are classified as type B; they partially include direct measurements. If the ESD model is based on the assumption that the predictors represent the true large-scale atmospheric state (analogous to the Perfect Prog approach, *Klein and Glahn*, 1974), type C variables should not be included. However MOS approaches (model output statistics, *Klein et al.*, 1959) also apply type C variables as the predictors (*Widmann et al.*, 2003). In this study, we use type A and type B variables as predictors (presented later in this section). Condition (iii) implies that the predictor includes climatic trends or variability changes that affect the predictand; otherwise these changes cannot be captured by the ESD model. This postulation is linked to the stationarity assumption, which is a major uncertainty in ESD (e.g., *Benestad et al.*, 2008). If the local predictand is subject to temporal changes not present in the predictors, the ESD transfer function becomes less appropriate with time. Several studies suggest to use multiple variable types as predictors to account for each changing variable that affects the local climate (e.g., *Fowler et al.*, 2007; *Hewitson and Crane*, 2006). In this study, both single- and mixed-field predictors are applied (see section 2.4.1).

Apart from the choice of variable types as predictors, ESD results also depend on the definition of the downscaling domain (i.e. the grid point area of the variable fields included in the ESD model) (*Fowler et al.*, 2007). The optimum choice of the predictor domain is generally not limited to the closest grid points around the study site, but includes the most important synoptic patterns around and upstream of the study area. Unrelated atmospheric variability, though, has shown to negatively impact the results in ESD (e.g., *Benestad et al.*, 2008). In terms of the climatic situation, the Cordillera Blanca is located in the outer tropics (e.g., *Kaser et al.*, 1996; *Kaser and Osmaston*, 2002), and the large-scale flow is dominated by easterly wind directions all year round (*Georges*, 2005). Anomalies of the large-scale flow, related to the strength and location of the Bolivian High, are associated with precipitation variability in the Andes (e.g., *Vuille et al* . 2008). Near surface moisture in the tropical lowlands also plays an important role for precipitation mechanisms (e.g., *Garreaud et al.*, 2003; *Vuille and Keimig*, 2004).

Since the Cordillera Blanca is a very complex mountain range, it is important to consider the NCEP model topography. Figure 2.2 shows the location of the glacier Artesonraju within the NCEP model topography: it is located east of the mountain crest in the model, whereas in reality the glacier faces mainly towards west and ends up in high peaks in the north and east. The four grid points closest to the study site are located between 1000 and 2000 m a.s.l., whereas the measurement site (AWS) is located at 4850 m a.s.l., with surrounding peaks above 6000 m a.s.l. Thus, the NCEP topography considerably differs from the real orographic situation of the study site and therefore, at sub-daily time scales, the direct use of grid point variables from NCEP/NCAR reanalyses for impact studies is not deemed adequate.

For all variable fields, we define first a preliminary horizontal grid point domain that extends from 17.5°N to 35°S and 102.5 to 50°W (entire area in figure 2.2). Due to the coarse NCEP model topography, the grid points at the model surface are not located in the pressure level of the study site. Consequently, the vertical allocation of the study site in the NCEP model is undetermined. In this study, we therefore consider both variable fields in the lower troposphere (1000 and 800 hPa), and the mid troposphere (600 and 400 hPa). Mid tropospheric variables have the advantage to be more reliable in the NCEP/NCAR reanalysis data (type A), whereas surface variables are influenced both by the model and by measurements (type B) (*Kalnay et al.*, 1996). We define the definite downscaling domain in this study by means of correlation maps between the preliminary predictor fields and the predictands, to identify regions with high correlation. From a theoretical perspective, this is suggested by the similarity of principal component-multiple regression analysis (the technique applied here, introduced later) with one dimensional maximum covariance analysis.

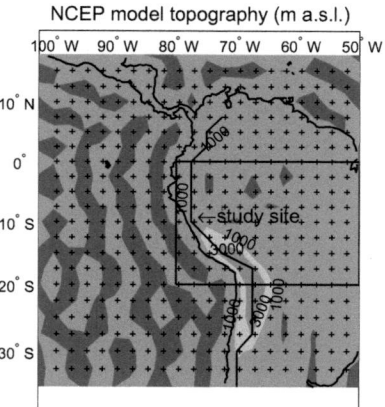

Figure 2.2: NCEP model topography over South America. The crosses are grid points. The black line connects the highest locations in the NCEP model topography. The black rectangle shows the horizontal area finally included in the ESD model.

The latter uses the regression map (which has a spatial structure very similar to the correlation map) to define the weights of the local predictors (e.g., *Widmann*, 2005).

In this study we select a preliminary predictor set from the NCEP/NCAR reanalysis data, including the variables air temperature a, specific humidity s, the horizontal wind components u and v, and the geo-potential height h. The variables a and s are the first candidates to be included in the multiple predictor set, because the physical linkage to the local predictands is most obvious for a and s, and they are important variables with regard to climate change (e.g., *Trenberth et al.*, 2007). We also include a set of circulation-related predictors (u, v, and h), which are more frequently applied variable types in ESD (e.g., *Cavazos and Hewitson*, 2005; *Fowler et al.*, 2007); not only because they explain a large proportion of variance in the predictands, but also because of the long temporal records, and the reliability of GCMs in simulating these fields. Starting from this preliminary predictor set, we try to find out the most appropriate fields to be included in the ESD model based on multiple regression analyses. The results of the downscaling domain and predictor variable selection are presented in section 2.4.1.

2.3.3 Transfer functions for individual times of day

In the outer-tropical Cordillera Blanca, seasonality is characterized by variations in atmospheric moisture rather than by temperature variations. The core wet season is defined from January to March, the core dry season from June to August, and the two transitional seasons span April to May and September to December (definitions based on precipitation climatology; *Niedertscheider*, 1990). It is important to note that the seasons are not of equal length over the year: this is a typical behavior of outer-tropical climates and is governed by the oscillation of the Inter-tropical Convergence Zone (e.g., *Kaser and Osmaston*, 2002).

To account for seasonal (and diurnal) changes, ESD variables are generally standardized individually for each month (and time of day, in the case of ESD in sub-daily time scales) (e.g., *Kim et al.*, 1984; *Schoof and Pryor*, 2001; *Xoplaki et al.*, 2003). In this study statistical transfer functions are derived separately for distinct months and times of day, because we assume that physical processes determining micro-meteorological conditions change for different months and times of day. In particular, diurnal air temperature maxima above temperate ice surfaces generally show small variations due to the stabilizing effect of the ice underneath, which cannot exceed 0°C. By contrast, night time air temperatures exhibit high variability, due to the high thermal emissivity of ice (e.g., *Paterson*, 1994). By applying separate transfer functions for different months and times of day, we include changes of the relationships between local and large-scale variables, as well as changes in means and variances of the variables.

Linear transfer functions are derived for each month of the year and the four times of the day when reanalysis data are available: 6, 12, 18, and 0 UTC (corresponding to 1, 7, 19, and 13 hours Peruvian Local Time, LT); hereafter referred to as "month/hour-models" (here we use the term "hour" instead of "time of day" for brevity; please note that "hour" does not refer to "hourly mean", but to instantaneous time of day, as defined above). This way, the calibration data set from AWS (March 2004 to May 2006) is split up into 48 sub-samples (corresponding to the 48 month/hour-models) of sample sizes n; each of them including approximately 30 to 70 observations (data gaps excluded).

2.3.4 Empirical orthogonal function (EOF) analysis and predictor pre-processing

Empirical orthogonal function (EOF) analysis is a convenient multivariate technique used in earth sciences to identify dominant variation patterns of random variables (e.g., *Hannachi et al.*, 2008; *Von Storch and Zwiers*, 2001). Various forms of EOFs are commonly applied in ESD for dimension reduction and to eliminate co-linearity in the predictors (e.g., *Benestad et al.*, 2008). The suitability of ESD techniques based on EOF analysis and related techniques (such as canonical correlation analysis for multivariate predictands; see *Von Storch and Zwiers*, 2001) has been reported in several studies (e.g., *Benestad*, 2001; *Hertig and Jacobeit*, 2008). In this study, we use the common terminology EOFs for the spatial, and PCs (principal components) for the temporal patterns, respectively, e.g., as proposed by *Hannachi et al.* (2008).

We employ EOF analysis for dimensionality reduction in the predictors (the definite predictor fields used are presented later, section 2.4.1). The predictors' time series over the entire forecasting period (here 1960 to 2008) are divided into 48 subsamples corresponding to the 48 month/times of day (as defined before). The 48 subsamples are transformed individually into single- and mixed-field (*Kutzbach*, 1967) EOFs. By applying EOF analysis separately for different months and times of day, the corresponding PCs do not contain annual or diurnal cycles any more, but anomalies thereof. Mixed-field EOFs are applied for the predictors containing multiple variable fields (because co-linearity appears not only spatially, but also amongst the different variable types). Before the calculation of EOFs, the variables are converted into dimensionless z-scores. The z-transformation is a step of data pre-processing especially important in the case of mixed-field EOFs, in order to remove the different physical units of the variables. In this study, the variables at each grid point are normalized individually (there is also the option of removing the field variance for each variable type). Geographical weighting is applied to the horizontal fields to account for increasing grid point density with latitude.

2.3.5 PC screening based on cross-validation

Once the selected predictors are prepared and transformed into EOFs and PCs, respectively, the number of PCs to be finally included as predictors for the month/hour-models needs to be determined (i.e., PC truncation and screening; e.g., *Wilks*, 2006). In many studies, a priori screening (or pre-screening) is applied, according to the percentage of explained variance of PCs in the predictor data (i.e., an a priori determined number of leading PCs are selected). However such choices are subjective. In this study, the PC selection is based on cross-validatory estimates of model error variances.

The procedure applied in this study is moving-blocks cross-validation (e.g., *Wilks*, 1997, 2006). The sub-samples for each month/hour-model are further divided into $m = n - L + 1$ of consecutive observations of length L (n is the number of observations of each month/hour-model). Ordinary least squares regression is then repeated m times to derive the predictor-predictand relationships by omitting each sub-sample once. The number of omitted observations L is determined for each month/hour-model based on the autocorrelation functions of the predictand time series, such that $L = 2(dct - 1) + 1$, where dct (decorrelation time) is the number of lags for which the serial correlation in the data is close to zero. Note that L has to be determined individually for each month/hour-model, because dct varies for the different sub-samples.

We define the mean square of the m differences between the m developed models and corresponding central withheld data points as forecast mean squared error, mse_f. By repeating the cross-validation procedure for different numbers of included PCs, k, and computing mse_f as a function of k, the optimum k can be determined. As model selection criterion we use the Akaike Information Criterion (AIC, *Akaike*, 1973), given by:

$$AIC(k) = ln(mse_f) + 2 \cdot \frac{n}{k} \qquad (2.2)$$

n is the number of observations. The optimum number k of PCs to be included, k_{opt}, is found at the minimum of $AIC(k)$. k_{opt} is determined individually for each month/hour-model. The use of $AIC(k)$ rather than $mse_f(k)$ for model selection has the advantage that for a given value of mse_f, the more parsimonious model is selected (i.e. the model with less predictors).

The ESD model (cf. equation 2.1) can be expressed now in terms of the month/hour (md)-transfer functions, such that:

$$Y_{md}(t) = \alpha_{md} + \sum_{npc} \left(\beta_{md}^{npc} \cdot PC_{md}^{npc} \right) + \epsilon_{md}(t) \qquad (2.3)$$

where Y is the predictand (air temperature or specific humidity) as a function of time t, α and β are intercept and coefficients of the linear multiple fit for each PC of the respective predictors (single- or mixed-field), npc is the order of included PCs, reaching from 1 to k_{opt}.

2.3.6 Model calibration and double cross-validation

The transfer functions (equation 2.3) are derived by least-squares regression. We use cross-validation to estimate the forecast skill of the regression equations. The data are resampled as described in section 2.3.5. The model fitting procedure, including the determination of k_{opt} based on cross-validation, is then repeated for each sample, i.e. by double cross-validation (e.g., *Michaelsen*, 1987). The regression coefficients (and uncertainties) are determined as the means (and variances) of coefficients from each repetition of the model fitting procedure in the double cross-validation experiment.

Again as independent goodness-of-fit measure, for each month/hour-model mse_{f} is calculated based on data withheld from the entire modeling procedure (including the cross-validatory PC screening). We define the skill score (SS) as:

$$SS = 1 - \frac{mse_{\text{f}}}{mse_{\text{rf}}}, \qquad (2.4)$$

where mse_{rf} is the mse obtained by fitting the mean of each month/time of day to the values. The subscript f (forecast) in mse_{rf} refers to the estimation of mean values using the same cross-validation procedure as for mse_{f}. SS can be interpreted as the added value over fitting the values to a reference model, and is used to evaluate the month/hour-models in this study, with the reference model being the mean value of the respective month/time of day estimated by cross-validation. SS is also known as reduction of error (e.g., *Wilks*, 2006).

2.4 Results and discussion

2.4.1 Predictor analysis

We use correlation maps to identify regions with high correlation between predictors and predictands (introduced in section 2.3.2). The results reveal maximum values above the Bolivian Altiplano of the reanalysis model, the southernmost parts of the selected domain, and Brazil (results not shown). The Bolivian Altiplano, because of its larger horizontal extent, is represented in the NCEP topography more realistically than the Cordillera Blanca (in terms of surface elevation). Consequently, the model climate above the Bolivian Altiplano also fits better to the conditions in the Cordillera Blanca, even if located about 10° further south. We reduce the preliminary downscaling domain (as defined in section 2.3.2) to cover an area from 80° to 50°W and 0° to 20°S (black rectangle in figure 2.2). The rectangular domain is not symmetric around the study area, but shifted upstream; towards the east. This way the area covers the significant synoptical patterns around, and upstream of the study site (section 2.3.2): the Bolivian High, and the Inter-Tropical Convergence Zone (e.g., *Garreaud et al.*, 2003; *Kaser and Osmaston*, 2002).

Figure 2.3 shows the correlation of the examined predictor fields a, s, h, u, and v, in the vertical levels 1000, 800, 600 and 400hPa, with the target variables air temperature and specific humidity. r^2 is the adjusted coefficient of determination obtained by multiple linear regression between the predictands and the first 10 PCs of each predictor field (X). Based on figure 2.3, the relative performances of the predictors can be evaluated. In the following discussion, a_{1000} denotes the variable a (air temperature) in the 1000 hPa level, and so on. The predictand air temperature generally shows higher correlations to the large scale predictors (> 0.5) than specific humidity (< 0.5). The best five predictors for air temperature, in terms of r^2, are a_{1000}, a_{800}, s_{1000}, u_{1000}, and u_{800} (in that order). The differences in r^2 of the predictor a_{1000} and other best performing predictors (e.g., s_{1000}) are small, which indicates considerable co-variability amongst these predictor fields. The variable fields in the lower troposphere generally show higher correlations than the predictors from the mid troposphere (possibly because of the larger diurnal variations in the lower troposphere that represent a major part in the predictand variances as well). For specific humidity, the predictor fields s_{1000} and s_{800} are the best predictors (out of the examined) with values of r^2 exceeding 0.4. Regression analyses performed to test all possible combinations of two and three predictors out of the examined fields (the 10 leading PCs of the mixed-field EOFs, respectively) reveal several variable combinations with comparable high correlations r^2 with the predictands.

Synthesizing the exploratory analyses above and the climatological criteria discussed in section 2.3.2, the pool of most appropriate predictors is more constrained, but the definitive choice remains ambiguous. In this study we repeat the modeling procedure (sections 2.3.5 to 2.3.6) with two predictors for each predictand: a_{1000} for the predictand air temperature, s_{1000} for the predictand specific humidity, and the combined field of a_{1000}, s_{1000} and u_{400} for both predictands. This way we want to examine the relative merits, as far as possible, of using multiple predictors against single-field predictors in ESD. Even though it does not show to be an important single predictor in figure 2.3, u_{400} shows the best performance as third predictor. The influ-

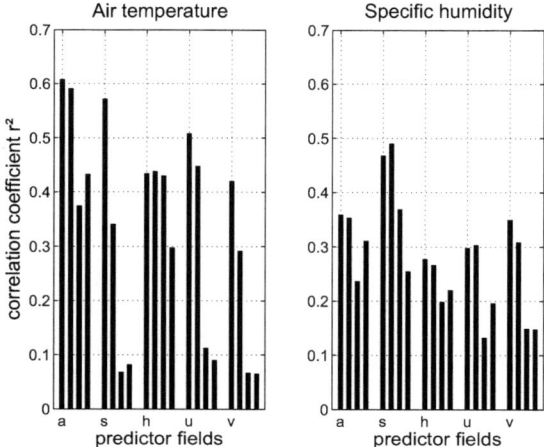

Figure 2.3: Adjusted correlation of determination (r^2) between the leading 10 PCs of the fields a, s, h, u, and v, for the predictands air temperature and specific humidity. The four bars of each variable correspond to the vertical locations of the predictor fields: 1000, 800, 600 and 400hPa levels from left to right.

ence of mid and upper tropospheric zonal winds to the local-scale variability in the Altiplano is described e.g. by *Garreaud et al.* (2003), who state that the winds trigger the lower tropospheric moisture transport by downward mixing of zonal momentum. Note that we do not use different predictor variable types or domains for the different month/hour-models.

2.4.2 Discussion of the month/hour-models

In this study, autocorrelation is taken into account by defining the block length of withheld samples in the cross-validation procedure, L, as described in section 2.3.5. Most of the month/hour-sub-samples (more than 90% of all cases) are serially uncorrelated for time lags (dct) of 1 to 4. The lag-1 autocorrelation is smaller than 0.5 (0.6) for air temperature (specific humidity) for more than 75% of the 48 month/hour-sub-samples. L is chosen to range from 1 to 7 in more than 90% of all cases. Note that for $L = 1$, the time series is not serially correlated and the moving-blocks cross-validation reduces to leave-one-out cross-validation. The resulting m (i. e., the number of independent residuals in the double cross-validation for the mse_f and mse_rf statistics) ranges from 20 to 60 and is larger than 40 in more than 75% of all cases (data gaps

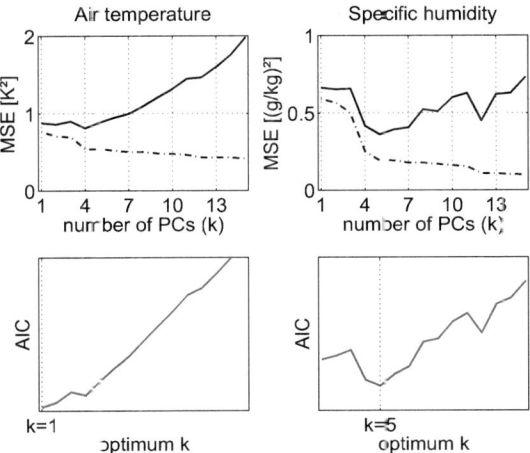

Figure 2.4: Illustration of the PC screening procedure applied in this study using the example of the monthly/hourly case December/13LT, for the predictands air temperature (left) and specific humidity (right) and the single-field predictor PCs of a_{1000} and s_{1000}, respectively. Top panel: mse_f (solid lines) and mse_h (dashed lines) as a function of k. Lower panel: $AIC(k)$ (arbitrary units).

considered).

Figure 2.4 is an illustration of the PC screening procedure applied in this study. mse_f and mse_h (i.e. the hindcast mse, defined as mse derived from the developmental sample) are plotted against the number of included PCs, k. In the experimental set up of this study, the number of predictors, k, means that PCs of the orders 1 to k are included as predictors in the multiple regression. As an example the case of December/13LT is shown, for both predictands air temperature and specific humidity and corresponding single-field predictors (a_{1000} and s_{1000}, respectively). While mse_h generally decreases as k increases, minimum mse_f is found at $k = 4$ (air temperature) and at $k = 5$ (specific humidity). In the example for air temperature, the minimum AIC is found at $k = 1$. Thus in this case AIC is the more parsimonious selection criterion than mse_f and the PCs 1 (and 1 to 5) are included as predictors in the December/13LT transfer functions for air temperature (and specific humidity).

The entire double cross-validation procedure, including the above illustrated PC screening, is repeated 192 times for the determination of transfer functions for each of the 48 months/times of day of all four predictand-predictor combinations (single-

and mixed-field predictors for the two predictands, cf. equation 2.2). Here we want to give an overview about the resulting functional relationships. About 50% of all relationships include 4 or less PCs (k_{opt} smaller than 4), and about 25% include 7 PCs or more. In particular, the more parsimonious models are found for predictand air temperature and for the single-field predictors, than for specific humidity or mixed-field EOFs. We refer to the absolute values of regression coefficients for the standardized predictors as weights. The weights of selected PCs are smaller than 0.2K in about 60% (single-field) and 90% (mixed-field predictors) of all month/hour-models of the predictand air temperature, and smaller than 0.2g/kg in about 60% (single-field) and 97% (mixed field predictor) of all month/hour-models of specific humidity. Relative uncertainties of the regression coefficients are defined here as standard deviations divided by mean values of the coefficients estimated in the double cross-validation procedure, section 2.3.6. The resulting uncertainties of about 90% of all estimated coefficients are smaller than 0.1%. In more than 70% of all cases the respective uncertainties of the estimated intercepts are smaller than 0.1% (for both predictands air temperature and specific humidity), and the uncertainties are slightly smaller in the case of mixed-field predictors, than single-field predictors.

2.4.3 Cross-validation of the month/hour-models

The month/hour-models are tested by moving-blocks cross-validation to evaluate the forecast skill (i.e., based on data not used in the calibration process). Skill scores (SS) are calculated based on mse_f as defined in section 2.3.6. As mentioned earlier, the SS can be interpreted as reduction of error against (in this experiment) the mean of the respective month/hour, estimated by cross-validation. The resulting values of SS are displayed in a normal probability plot (figure 2.5), to facilitate the discussion of all 48 month/hour-models for each of the four predictor/predictand combinations.

For the combination air temperature and single-field predictor (a_{1000}), more than 60% of the month/hour-models SS is smaller than 0.2 (with even slightly negative values), indicating no or only few added value of the respective transfer function over a constant value. Some regression functions also show high values of SS (larger than 0.6). For the mixed-field predictors, however, the respective values are clearly more positive (cf. figure 2.5): 50% higher than 0.3, 25% higher than 0.5, and maximum values reaching up to 0.8. The combination specific humidity with mixed-field predictor also shows higher SSs, than with the single-field predictor (figure 2.5); e.g., about 65% are higher than 0.2 for the single-, and 75% for the mixed-field predictors.

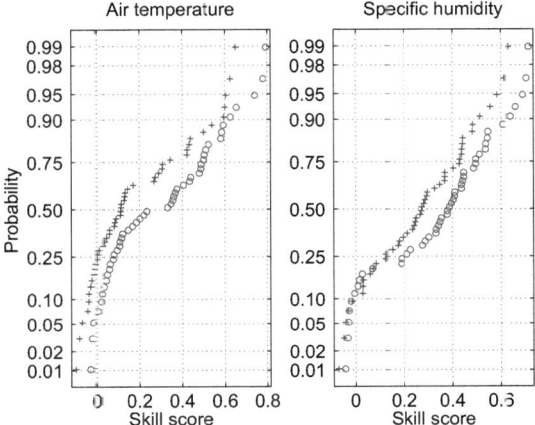

Figure 2.5: Skill scores of the month/hour-transfer functions estimated by double cross-validation for the predictand air temperature (left) and specific humidity (right), single-field (crosses) and mixed-field (circles) predictors.

Generally, there are less SSs smaller than 0.2 for the predictand specific humidity than for air temperature.

To summarize, the forecast skill of the developed transfer functions quantified by SS varies from 0 to 0.8, but is higher than 0.3 in more than 50% of the cases, and the mixed-field predictors show generally higher skill than the single-field predictors for both predictands. $SS = 0$ (as is the case for some month/hour-models) means that the deterministic part in equation 2.1 is zero. In this case, the resulting model reduces to the mean of the respective case. In the next section, we want to give a more comprehensive discussion on how the varying skill of the different month/hour-models affects the ESD model as complete time series.

2.4.4 Performance of the ESD model against simpler reference models

The modeled time series of the individual month/hour-models are put together to form a complete time series over the entire forecasting period 1960 to 2008 (hereafter referred to as ESD model). Note that in this section, unlike the cross-validation experiment in section 2.4.3, the performance of the ESD model is assessed based on data used in the calibration process (i.e. hindcast performance). We define reference model 1 (RM1) as the mean diurnal cycles of air temperature and specific humidity for each month of the year, calculated from the AWS observations. Since the predictor PCs in the month/hour-models are standardized, RM1 is equal to the respective intercepts of the month/hour-transfer functions (equation 2.3). Note that RM1 is not a climatology, given the short averaging period (2004-2006).

Figure 2.6 shows a case study of 6-hourly observations, RM1, and the ESD model in April 2004. The example shows a period where melting at the glacier surface occurred (e.g., *Mölg et al.*, 2009). Within the first few days, night-time air temperatures above 0°C and high values of specific humidity (up to 6g/kg) indicate that increased longwave incoming radiation constrained the cooling of the ice surface and thus air temperature variability. In the second half of the case study, melt intervals were shorter and air temperature varied more strongly due to lower minima. The linkage between melt occurrence and atmospheric moisture in the tropics is well known: if humidity is low, sublimation consumes much energy and reduces or prevents melting (e.g., *Kaser and Osmaston*, 2002). In fact, specific humidity is clearly lower in the second half of the case study period. The ESD model is able to simulate the small diurnal cycles of air temperature associated with high values of specific humidity in the first few days, and the drop in humidity accompanied by larger diurnal cycles of air temperature in the second half of the case study. This case study also exemplifies how high-resolution air temperature and specific humidity are linked to glacier melt, which can be simulated in energy balance-based mass balance models (*Mölg et al.*, 2009). RM1, defined as constant diurnal cycle for each month, cannot show the drop in specific humidity on 04/16. In this context it is important to see that the ESD model (which per definition reduces to RM1 when the SSs of all the month/hour-models are zero) captures the above described day-to-day variability that significantly affects glacier mass balance.

In order to quantify the added value of the ESD model versus the reference model over the entire calibration period, the hindcast skill score (SS_h) is defined as:

$$SS = 1 - \frac{mse_{h,ESD}}{mse_{h,ref}}, \quad (2.5)$$

where $mse_{h,ESD}$ and $mse_{h,ref}$ are the hindcast mean squared errors of the ESD model and the reference model, respectively (i.e., the mean squared differences to the observations over the entire calibration period). The values of SS_h are shown in table 2.1. In the hourly time scale, values of SS_h range from 0.2 to 0.4. SS_h is higher for specific humidity than for air temperature, and higher in the case of mixed-field predictors than for single-field predictors.

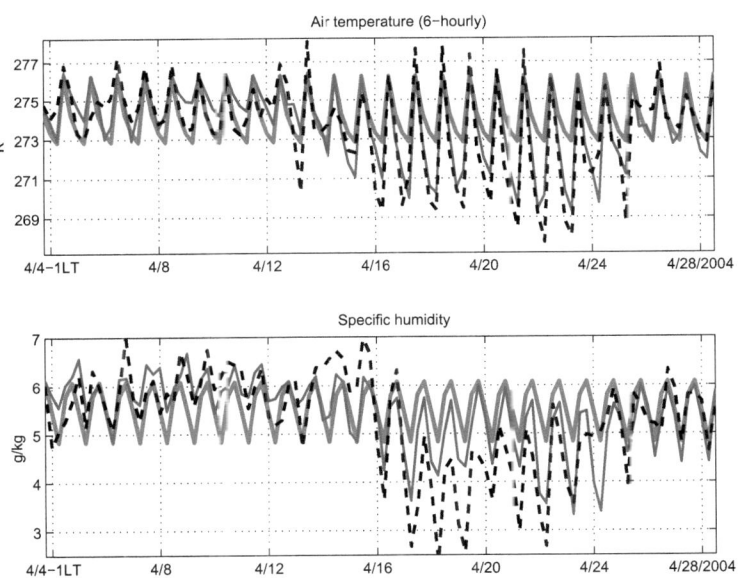

Figure 2.6: Diurnal cycles of air temperature (top) and specific humidity (bottom) at AWS: observations (dashed), ESD model hindcast (thin), and reference model (solid); a case study in April 2004 (1LT in the first x-tick means 1 Peruvian Local Time).

Table 2.1: Hindcast skill scores SS_h of the ESD model (mixed-field predictors) against reference models RM1 and RM2 (equation 2.5); for the two predictands air temperature and specific humidity and different temporal scales.

	air temperature			specific humidity		
	sub-daily	daily	monthly	sub-daily	daily	monthly
RM1	0.37	0.47	0.63	0.43	0.56	0.57
RM2	0.75	0.63	0.93	0.67	0.71	0.79

To evaluate the hindcast performance of the ESD model for different temporal scales, daily and monthly means are built from the 6-hourly observations and ESD model in the calibration period. The respective reference models in the daily time scale is reduced to seasonal cycles represented by monthly means of the observations. Figure 2.7 shows observations, reference model, and ESD model daily means as a function of time over the entire calibration period. The added value of the ESD model over the reference model is most evident in the daily time scale, since - per definition - in the reference model no day-to-day variations occur within one month. Values of SS_h are higher than in the hourly time scale, ranging from about 0.45 to 0.55 (see table 1).

In the monthly time scale, the ESD model fits the observations almost perfectly (figure 2.8). The resulting values of SS_h show high variations for the four predictand-predictor combinations, ranging from 0.32 (for air temperature/single-field predictor) up to 0.93 (for specific humidity/single-field predictor). SS_h is a relative measure of skill and must be considered with care. In the monthly time scale in particular, both the ESD model and the reference model fit the observations very closely, because the calibration period is so short (12 transfer functions and monthly means based on about 24 months of data). Consequently the values of mse are very low (< 0.1) and minor changes in mse result in high variations of SS_h.

Figure 2.8 also shows monthly means of NCEP/NCAR reanalysis data from the nearby grid point at 500hPa, 77.5°W and 10°S (note that this grid point is not the closest of the eight grid points surrounding the study site in three dimensions, but the one that shows the highest correlation to the observations at AWS, as has been revealed in the study of *Hofer*, 2007). The curve representing the NCEP grid point values is shifted by its mean bias to the observations (given the 500 hPa level is located far from the NCEP model surface, temperature and humidity values are systematically too low). NCEP grid point values are not shown in figures 2.6 and

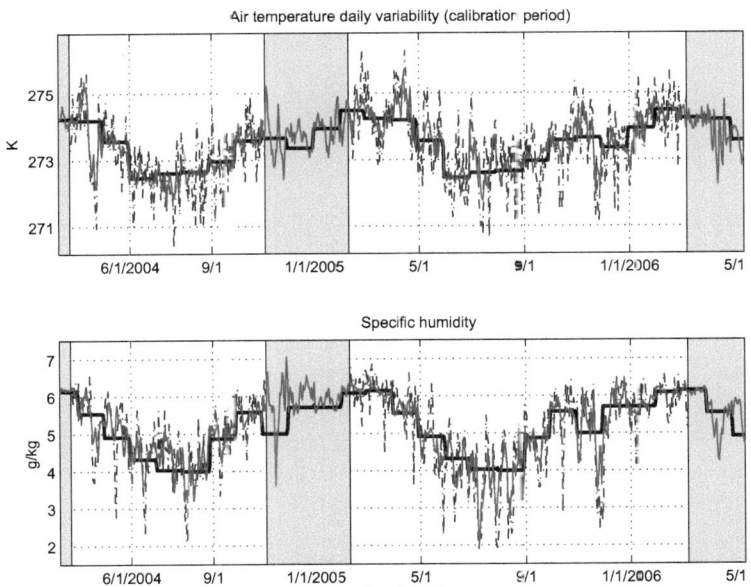

Figure 2.7: Daily mean air temperature (top) and specific humidity (bottom) at AWS: observations (dashed), ESD model hindcast (thin), and reference model (solid) over the calibration period (March 2004 to May 2006). Periods with missing data are grey shaded.

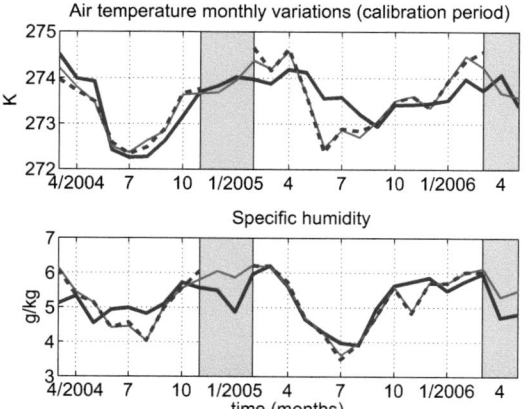

Figure 2.8: Monthly means of air temperature (top) and specific humidity (bottom): observations (dashed), ESD model hindcast (thin), and reanalysis grid point (solid) in the calibration period (March 2004 to May 2006). Periods with missing data are grey shaded.

2.7 (6-hourly and daily values) in order to keep the figures simple, however, there is much less agreement to the observations than in the monthly timescale. To evaluate the added value of the ESD model hindcast (using mixed-field predictors) against the simple use of grid point values, we define SS_h similarly as in equation 2.5, but the reference model is represented by the NCEP grid point values, hereafter referred to as RM2. Hence $mse_{h,ref}$ is the mse between RM2 and the observations. Since RM2 has zero bias by construction in this study, $mse_{h,ref}$ is an estimate for the error variance. SS_h is 0.75 (6-hourly), 0.63 (daily), and 0.93 (monthly) for the predictand air temperature, and to 0.67 (6-hourly), 0.71 (daily), and 0.79 (monthly) for specific humidity. Even if the NCEP grid point values fit the seasonal cycle already closely (cf. figure 2.7), values of SS_h in the monthly time scale are still very high because $mse_{h,ESD}$ is close to zero. Overall, SS_h is higher than 0.6 for both predictands at all time scales.

2.4.5 ESD model forecasts 1960 to 2008

The training period of the developed ESD model (calibration and validation) represents less than 3 years; about 5% of the reanalysis period. The question arises whether the ESD model is valid outside the training period. In particular, is the model able to reproduce variations that it has not experienced within the training period? If the deterministic part in equation 2.1 is zero, the ESD model is equal to the RM1 and consists of constant values for each calendar month, and time of day. In this case the inter-annual variations are zero. If the random part in equation 2.1 is zero, the ESD model fits the observations perfectly. The double cross-validation procedure yields skill scores SS ranging from 0 to 0.8 (figure 2.5), indicating that the random part in equation 2.1 for some months/times of day is large and the resulting transfer functions are constant (or almost constant). Consequently, we can expect the ESD model to underestimate the variability at inter-annual time scales (similarly as it does in the time scale of model fit, see *Von Storch*, 1999).

Inter-annual atmospheric variability in the Cordillera Blanca is mainly governed by the El Niño Southern Oscillation (ENSO); positive temperature anomalies are connected with El Niño and negative ones with La Niña events. Concerning atmospheric moisture, El Niño years often tend to be dry and La Niña years tend to be wet, but the reverse is not uncommon as well (e.g., *Garreaud et al.*, 2003; *Vuille et al.*, 2008). Figure 2.9 (top) shows annual means of the ESD model forecast (single-field predictors) 1960 to 2008 for both predictands (air temperature and specific humidity). As a reference, annual means of the NCEP grid point values (77.5°W/10°S in 500hPa of both air temperature and specific humidity) and the observational time series from Querococha (only air temperature, from 1965 to 1994) are displayed (centered values). Note that these time series, even if expected to be similar, must not be identical (i.e., differences do not necessarily come from errors in the ESD model).

For air temperature (figure 2.9 top left) ESD model, NCEP grid point and observational time series generally show similar patterns that clearly reflect ENSO. While the NCEP grid point time series shows a rising trend over the period, no such trend is evident in the Querococha time series and the ESD model. A comparison of ESD model forecast to the annual means in the calibration period (grey shaded area in figure 2.9) shows larger variability in the forecast period than in the calibration period, supporting the stationarity of the ESD model. However, inter-annual variability is smaller in the ESD model than in the NCEP grid point and Querococha time series. As discussed above, the ESD model underestimates the real variability by construction. However, the permanent exchange of latent heat due to sublimation, melting and refreezing smoothes air temperature variations above a glacier surface at all time scales and could add to the smaller variance in the ESD model. Specific humidity variances (figure 2.9 right top), by contrast, show similar magnitudes in both the ESD model and the NCEP grid point time series, but the temporal patterns over the

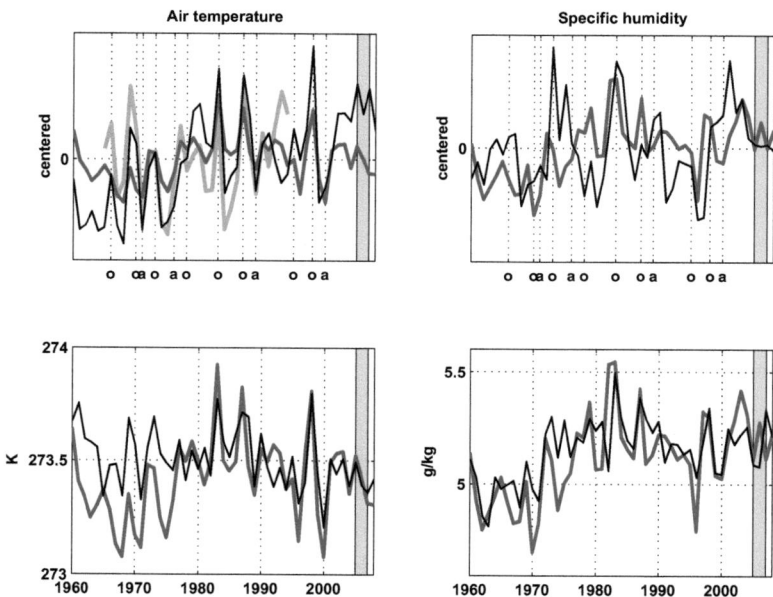

Figure 2.9: Top: Annual means of ESD model forecast (dark), NCEP grid point values (thin) and observations at Querococha (solid grey, only air temperature) of air temperature (left) and specific humidity (right) for the period 1960 to 2008. El Niño (o) and La Niña (a) events are indicated in the abscissa. Bottom: ESD model forecast of annual mean air temperature (left) and specific humidity (right) for the period 1960 to 2008, with two different predictor fields applied for each variable: the single-field predictors a_{1000} for air temperature and s_{1000} for specific humidity (solid), and the mixed-field predictors including a_{1000}, s_{1000}, and u_{400} (black). The model training period (2004 to 2006) is grey shaded.

years differ. Also, the relation of specific humidity to El Niño/La Niña events is less evident in both time series, and sometimes reverses.

Figure 2.9 (bottom) again shows the ESD model forecast 1960 to 2008 for air temperature and specific humidity, but this time with two different predictors (single- and mixed-field). We want to assess if the chosen model is sensitive to the variable fields used as predictors. In fact, the two time series shows different patterns. For air temperature, the models are very similar in recent years, but prior to the late seventies the two solutions diverge. Still, they show more or less the same pattern of maxima and minima. For specific humidity, the ESD model based on mixed-field predictors shows less inter-annual variability than the ESD model based on single-field predictors, but the general pattern over time is similar (e.g., both time series show an increasing trend until the early 1980s and a slight decrease thereafter).

2.5 Conclusions

We have presented a computationally cheap method to transfer coarse-scale re-analysis data to local air temperature and specific humidity in a glacier environment at high temporal resolution. The developed non-parametric technique is appropriate for short time series and applicable wherever high-resolution observations are available. The modeling procedure includes an automated cross-validation scheme, but appropriate choices need to be made concerning the exact definition of predictor domain and variable fields. So far, we have not assessed the performance of NCEP/NCAR reanalysis data against other reanalysis products (e.g., *Kazutoshi et al.*, 2007; *Uppala et al.*, 2005). Also, EOF analysis is used in this study only for dimensionality reduction and not for pattern interpretation.

The developed high-resolution ESD model shows added value over simpler techniques, such as the use of reanalysis grid point values or constant annual/diurnal cycles calculated from the calibration data set. We are therefore confident that this study represents an important progress for physically-based mass balance studies on data-sparse tropical glaciers. As a next step, we schedule case study simulations using a regional atmospheric model to increase the knowledge about significant dynamic processes above a tropical glacier surface. By combining statistical and dynamical techniques, we target for the complete set of atmospheric key variables (including spatial gradients) required for long-term, process-based glacier mass balance modeling in the Cordillera Blanca.

CHAPTER III

Paper 2: Skill assessment of NCEP/NCAR reanalysis data for daily air temperature on a glaciated mountain range (Peru)

Marlis Hofer, Ben Marzeion, and Thomas Mölg

Faculty of Geo- and Atmospheric Sciences, University of Innsbruck (Austria)

(at the time of printing, since 05/2011 under review by *Journal of Climate*)

3.1 Introduction

This study presents a method that links reanalysis data to local measurements. The goal is to develop a comprehensible empirical-statistical downscaling (ESD) procedure that can easily be transferred to different sites, predictors, or predictands; and that includes a non-parametric estimate of the ESD model skill. The method is designed for predictor selection in the case of short (few years), sub-daily to daily time series of normally distributed target variables. In this study, the ESD method is used to quantify the skill of NCEP/NCAR reanalysis data for local-scale, daily air temperature measured at high-altitude automatic weather stations (AWSs) in the tropical Cordillera Blanca.

The Cordillera Blanca is a glaciated mountain range located in the Northern Andes of Peru (figure 3.1). It harbours 25% of all tropical glaciers with respect to surface area (e.g., *Kaser and Osmaston*, 2002). The glaciers are shrinking since their last maximum extent in the late 19th century (e.g., *Ames*, 1998; *Silverio and Jaquet*, 2005; *Georges*, 2004) and have significantly shaped the socio-economic developement in the region. Nowhere else in the world have glaciers caused such catastrophic disasters,

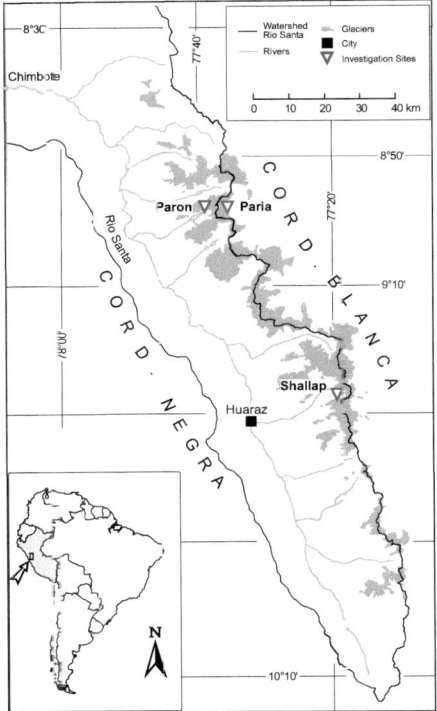

Figure 3.1: The map shows the Rio Santa watershed with the Cordillera Blanca mountain range and measurement sites (mentioned in the text). Also indicated is the 1990 glacier extend (grey shaded area) (*Georges*, 2004).

in terms of outburst floods and avalanches (*Carey*, 2005, 2010). Nevertheless the glaciers have important positive impacts for water availability in industry, agriculture and households because they contribute to balancing the high runoff seasonality in the extensively populated Rio Santa valley (e.g., *Juen*, 2006; *Juen et al.*, 2007; *Mark and Seltzer*, 2003; *Kaser et al.*, 2003, 2010). To relate future climate change to glacier shrinkage in the Cordillera Blanca, a better knowledge of past atmospheric and glacier variations is required.

More specifically, we ask how much information about recent-past, local-scale atmospheric variations can be gained from reanalysis data. Reanalysis data, similar to analysis data in numerical weather prediction (NWP), are a combination of general circulation model "first guess" and quality-controlled observations, and are generated using a data assimilation system. First proposed in the studies of *Bengtsson and Shukla* (1988), and *Trenberth and Olson* (1988) "re" -analyses have the advantage over NWP analyses that their production is based on a fixed modeling system for the entire assimilation period. Thus data discontinuities due to changes in atmospheric model and assimilation techniques are avoided. Today, global reanalysis data are available from four institutions worldwide: NCEP (*Kalnay et al.*, 1996), ECMWF (*Uppala et al.*, 2005), JMA (*Kazutoshi et al.*, 2007), and NASA (*Bosilovich*, 2008). They are extensively used in many fields of the atmospheric and oceanic sciences; in climate change and climate change impact studies, but also in predictability studies, for seasonal forecasting, and in dynamic system studies (e.g., *Rood and Bosilovich*, 2009; *Simmons et al.*, 2007).

The downscaling approach presented here is based on model output statistics (MOS) and does not rely on the perfect prog ("perfect prognosis", PP) assumption (*Klein et al.*, 1959). The difference between MOS and PP approaches is shown in figure 3.2. In PP the predictors in the calibration process are observations (PP-T in figure 3.2). We might also denote PP an approach that uses reanalysis data for model training, because reanalysis data are closely related to observations (PP*-T in figure 3.2). For the forecast the calibrated downscaling model is then applied to a GCM output by considering it as the true large-scale atmospheric state (PP assumption; PP, and PP* in figure 3.2). PP approaches thus make no attempt to correct for model biases or errors (e.g., *Wilks*, 2006). By contrast, MOS approaches (*Glahn and Lowry*, 1972) use output from the same GCM for both the statistical model set-up (MOS-T in figure 3.2), and forecast (MOS-F); thus accounting for GCM-specific biases and errors. The method presented here (number 3 in figure 3.2) is a MOS technique as it is trained by and applied to the same large-scale model (i.e. the NCEP reanalysis model, see below). In this study, by contrast to standard PP and MOS approaches (numbers 1, 2, and 4 in figure 3.2), the forecast period is in the past (number 3), because the goal is to extend the available short-term time series into the past.

Hofer et al. (2010) present an ESD model that links NCEP/NCAR reanalysis data to time series of air temperature and specific humidity above a glacier surface in the Cordillera Blanca at a sub-daily time scale. In their study, *Hofer et al.* (2010) use variable fields as predictors. The predictor selection includes synoptic considerations, but is primarily based on statistical data inference (using a double-cross-validation scheme). However, *Hofer et al.* (2010) point out ambiguity in the definite choice of input predictor fields. Here we present a method that for the first time provides a sys-

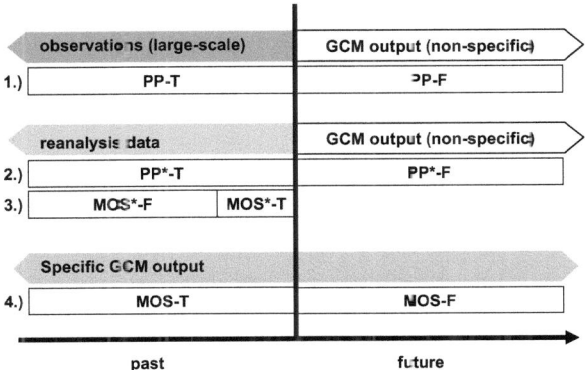

Figure 3.2: Schematic of the two classical statistical downscaling approaches MOS and PP, and options (*). The abbreviation T denotes training period, and F forecasting (or model application) period of the different approaches. The numberings of the different approaches 1-4 are referred to in the text.

tematic tool for predictor selection and comparison studies. We show an application of the method for daily air temperature predictands measured at several automatic weather stations in the Cordillera Blanca, and NCEP reanalysis data predictors. Note however that the presented ESD procedure can be applied to any kind of large-scale data predictors (e.g., different reanalysis products, gridded large-scale observations, or GCM output).

3.2 Study site and observations: the predictands

An observational network of several AWSs at and nearby glaciers in the Cordillera Blanca has been installed since 1999 (*Juen*, 2006), primarily to provide data for glacier mass balance modeling. Figure 3.1 shows the Cordillera Blanca with the investigation sites. The AWSs are located between 4700 and 5100 m asl at three main sites: (1) four AWSs at and near glacier Artesonraju in the Paron valley (Northern Cordillera Blanca), (2) one AWS in the Paria valley east of the main mountain crest at the same latitude as Artesonraju glacier, and (3) two AWSs in the Shallap valley at the west side of the main crest, about 100 km south of Artesonraju glacier. The AWSs are equipped to measure air temperature, relative humidity, air pressure, solar radiation, and wind speed and direction. Three AWSs located immediately on the glaciers (Artesonraju and Shallap) are measuring the full radiation balance (i.e. short- and longwave in- and outgoing radiation), two thereof only since recently (July 2010). Three AWSs located on moraines include precipitation measurements; one thereof (Paria) measures only precipitation.

Maintaining the AWSs to provide continuous and reliable atmospheric time series has been a logistical challenge. In technical terms, the most problematic measurements have been wind and radiation above the glacier surface (where it is difficult to guarantee leveling of the instruments), and precipitation, which are generally problematic in terms of snowfall and require more regular field visits. Besides the technical difficulties, problems include instrument theft and natural hazards (*Juen*, 2006).

The predictands in this study are daily air temperature time series measured at three AWSs located on moraines nearby the glaciers Artesonraju (hereafter referred to as AWS1 and AWS2) and Shallap (hereafter AWS3; the sites are indicated in figure 3.1). To date, air temperature time series of AWS1, AWS2, and AWS3 are the longest of all AWSs, i.e. five years from 2006/07 and to 2010/07 (including data gaps). AWS1 and AWS3 are located at similar high altitudes (5000 and 4950 m a.s.l, respectively), but approximately 100 km apart, whereas AWS1 and AWS2 (4800 m a.s.l.) are very close (approximately 1 km apart) but AWS2 is at a lower altitude. All three measurements are carried out with a HMP45 sensor by Väisalla and a ventilated radiation shield, described by *Georges* (2002).

Figure 3.3 shows statistics of the AWS1, AWS2, and AWS3 daily air temperature measurements for each month of the year. The measured time series are approximately normally distributed (not shown). The seasonal cycles in the data are small ($< 2°C$) and are characterized by two maxima and two minima throughout the year. At all AWSs the warmest months are January, April and November and the coldest March and July. The 25th and 75th percentiles shown in figure 3.3 indicate high within-month variabilities in January to March for data from AWS1 and AWS2, and in December to February for AWS3, which points to El Niño Southern Oscillation (ENSO) variability playing an important role in the region at this time of the year (e.g., *Vuille et al.*, 2008). Interestingly, the annual air temperature minimum is less pronounced at the southernmost located AWS3. Note, however, that these statistics should not be overvalued as a climatology, because they are based on only five years of measurements and thus the seasonal patterns are likely to be different for longer periods. Also shown in figure 3.3 are statistics for NCEP reanalysis data that will be referred to later.

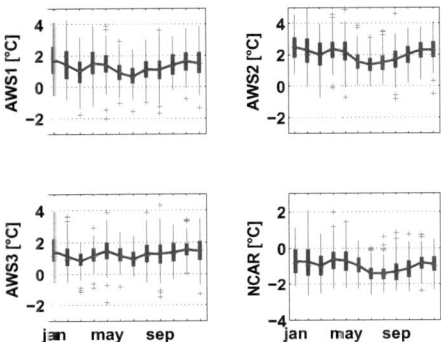

Figure 3.3: Statistics of hourly air temperature time series at AWS1 (5000 m a.s.l.), AWS2 (4800 m a.s.l.), and AWS3 (4950 m a.s.l.), and 6-hourly values of the NCAR reanalysis data predictor $air500600$ (as defined in the text), for each month of the year (abscissa: January to December). Shown are the means (blue solid line) and the medians (red dashes). The edges of the thick bars are the 25th and the 75th percentiles. The thin bars extend to the most extreme data not considered as outliers, and the crosses are the outliers. The statistics are computed over the period of available measurements July 2006 to July 2010.

In this study, even though not obvious from figure 3.3, we refer to a definition of climatological seasons in the Cordillera Blanca (dry, wet, and intermediate seasons), which are based on the study of *Niedertscheider* (1990) and also used in the study of *Juen* (2006). In short, the core humid season in the Cordillera Blanca is from January to March (more than 50% of the annual precipitation falls), and the core dry season from June to August (including less than 2% of the annual precipitation).

3.3 ESD model architecture

3.3.1 Accounting for seasonal periodicity

If atmospheric time series are considerably shorter than thirty years and the climatological seasonal cycle is not known, the problem arises how to strictly distinguish periodic, seasonal variations from aperiodic (or less periodic), day-to-day and interannual variability. Especially in statistical forecasting, periodicity must be accounted for to avoid that the periodic, seasonal variations dominate the model fit. When long enough data series are available, the problem is often avoided by subtracting the climatological seasonal cycle from the time series (e.g., *Madden*, 1976). This way seasonal periodicity is removed from the time series, but not necessarily from the model error.

In the present study we assume that seasonal atmospheric periodicity leads to changing relationships between large- and local-scale atmospheric variables throughout the year. Considering the atmospheric seasonal cycle in ESD models is important especially if the study site is located in the mountains. E.g., local-scale atmospheric conditions can be affected by topographic shading that changes with the solar altitude throughout the year, but the topography is misrepresented and thus these effects can not be captured by the large-scale model. Due to the same effects related to the diurnal cycle (sub-daily data) different models for the different times of day are required (e.g., *Hofer et al.*, 2010). By consequently using separate statistical predictor-predictand transfer functions for the different months of the year, seasonal periodicity is eliminated not only in the time series, but also in the model error. In practice in this study, each predictor-predictand pair is divided into twelve separate time series for each month, the number of observations in each time series consequently being approximately $n = N/12$, where N is the length of the complete data series. Then the modeling procedure is repeated identically for each calendar month's time series.

3.3.2 Predictor selection

We distinguish two different ways of predictor selection, namely (1) a priori predictor selection (based on knowledge outside the data, prior to looking at the data or to data analysis), and (2) data-based predictor selection. Most downscaling studies more or less systematically use a combination of both ways, by first pre-selecting a subset of potential predictors from an available pool (i.e., a priori selection), and then choosing the definite, final predictors based on criteria derived from the data (i.e., data-based selection; e.g. *Klein and Glahn*, 1974; *Wilby et al.*, 2002). However, few studies have systematically assessed the skill of different predictors in terms of variable types or grid points, and there is no consensus on the most appropriate choice. Here we present an a priori predictor selection in its strict sense because for data-based predictor comparisons, a priori selections are more useful than data-based selections.

What information of a large-scale atmospheric model would we use to represent local, daily air temperature if no observations were available? As simplest a priori predictor choice we relate the same physical predictor and target variables. For our purpose, this implies to use large-scale air temperature as predictor for local-scale air

Table 3.1: Specifications of the NCEP grid points applied as predictors: heights (h) and mean geopotential heights (gph) with standard deviations in brackets (all values are in units meters above see level).

	h [m asl]	gph 600 hPa [m asl]	gph 500 hPa [m asl]
285E 10S (SE)	1719	4408 (+/-11.3)	5867 (+/-13.5)
285E 7.5S (NW)	712	4408.6 (+/-11.7)	5866.8 (+/-12.8)
282.5E 10S (SE)	1993	4408.6 (+/-11.5)	5866.8 (+/-13.5)
282.5E 7.5S (NE)	1910	4409.9 (+/-11.9)	5867.3 (+/-12.9)

temperature. Note that this choice is problematic for variables that are related to large model uncertainty, such as precipitation; then it is generally recommended to use more accurately modeled and observed predictors (e.g., *Maraun et al.*, 2010) (however some MOS approaches exist that use precipitation as predictor; e.g., *Widmann et al.*, 2003).

Predictor selection includes not only the choice of a physical variable type, but also of geographical allocation in terms of model grid points. The optimum downscaling domain is generally not limited to the closest grid points around the study site, but includes important synoptic patterns around and upstream of the study area (*Benestad et al.*, 2008). For studies that use grid point predictors (e.g., *Brinkmann*, 2002) propose that not necessarily the closest grid points are the best predictors. Yet the choice of remote grid points or spatial patterns necessarily includes data-based selection and thus in this study, we restrict the choice to grid points nearby the study site. Since coarse-scale surface model topographies do generally not correspond to the real surface elevation of a mountainous site, the question arises whether (1) surface predictors (to account for important surface processes), or (2) predictors from the same elevation (or pressure level), are the more realistic choice for a predictand located at the surface. All three AWSs used in this study are located between 500 and 600 hPa at 5000 (AWS1), 4825 (AWS2), and 4950 (AWS3) m a.s.l. Table 3.1 shows coordinates, surface elevations, and geopotential heights for the 500 and 600 hPa levels of the four closest (relative to the study site) grid points in the NCEP reanalysis model. The grid points do not even exceed 2000 m a.s.l., thus it is evident the Cordillera Blanca is not represented realistically in the NCEP reanalysis model topography. There is very small horizontal variation in the mean 600 (and 500) geopotential heights between the four grid points, all located at approximately 4410 m a.s.l. (and 5870 m a.s.l., respectively, see table 3.1).

In numerical climate modeling, the skillful scale is the spatial domain, at which the models provide useful information (e.g., *Zorita and Von Storch*, 1997; *Von Storch et al.*, 1993; *Benestad et al.*, 2008). Whereas it differs largely for different models, the skillful scale (or optimum scale, *Räisänen and Ylhäisi*, 2011) is generally larger than the minimum scale (i.e. the distance between two neighboring grid points) because of numerical noise related to single grid point data (*Grotch and MacCracken*, 1991; *Willamson and Laprise*, 2000), usually comprising an area of several grid points. Accordingly in this study we propose to calculate a mean over several grid points to avoid this problem (see below).

We define air temperature averaged over the eight closest grid points (table 3.1, four horizontal locations at two vertical levels; i.e. the edges of the grid box that includes the study site) as a priori predictor for daily air temperature measured at the study sites, hereafter abbreviated with *air*500600. Concerning the exact choice of averaged grid points, this is only a starting point, and a systematic, data-based assessment is necessary in order to gain more information about the optimum domain. *air*500600 is proposed as a priory choice because it can be applied equally for different sites, seasons and large-scale models, without data inference. Note however that it may not necessarily be the best choice in each individual case. Even if we assume that the large-scale atmosphere is perfectly represented (PP assumption, i.e. no inter-model uncertainty), we expect a large-scale variable to represent the same local predictand not accurately. The coarser the large-scale model, the more complex we expect the relationship between the model and the local predictand to be, e. g. involving multiple large-scale variables, and other than linear relationships. In practice, the relation between large- and local-scale variables can best be investigated with a physically-based limited-area model.

Statistics of *air*500600 are shown in figure 3.3 together with the statistics of the AWS data for each month of the year over the calibration period.

3.3.3 Downscaling process: linear model calibration and cross-validation

In this section the entire ESD modeling procedure, including data preprocessing, ESD model calibration, and skill estimation based on cross-validation is presented. Cross-validation is important especially in the case of short-term observational time series (as in the present study), because it allows each observation to be used in the model building process as well as in the model evaluation process. The modification of leave-one-out cross-validation specifically presented here is appropriate for daily or sub-daily atmospheric time series, because it accounts for temporal autocorrelation (i.e., persistence, *Madden*, 1979).

First, the predictor and predictand time series (daily means) are separated into twelve different time series for the twelve months of the year. All steps described below are repeated separately and independently for each month's time series.

The simplest way to relate an a priori predictor to a Gaussian target variable is a linear regression model. Note that the model is not appropriate for non Gaussian target variables (e.g., precipitation). It applies

$$y_s(t) = a \cdot x_s(t) + \epsilon(t), \tag{3.1}$$

where a is the least-squares regression parameter, ϵ the model error, and $y_s(t)$ and $x_s(t)$ are the standardized predictand and predictor time series.

Let $\overline{\cdot}^t$ be the temporal mean, and $\sigma_t(\cdot)$ the temporal standard deviation of a variable (here the standardization parameters). Because $\overline{y_s}^t = \overline{x_s}^t := 0$ and $\sigma_t(y_s) = \sigma_t(x_s) := 1$, and with $\tilde{y}(t) := y(t) - \epsilon(t)$, equation 3.1 can generally be rewritten including untransformed predictand $y(t)$, predictor $x(t)$, and $\overline{\cdot}^t$, and $\sigma_t(\cdot)$:

$$\hat{y}(t) = \overline{y}^t + a \cdot \frac{\sigma_t(y)}{\sigma_t(x)} \left(x(t) - \overline{x}^t \right) \tag{3.2}$$

To estimate $\hat{y}(t)$ and $\hat{\epsilon}(t)$ a modification of leave-one-out cross-validation (*Michaelsen*, 1987) is applied. Cross-validation is repeated n_{cv} times. Here we use $n_{cv} = n$ (n is the number of observations of each month-separated time series). Each cross-validation repetition, n_{lo} observations are excluded from the model fit (the "left-out" observations), with

$$n_{lo} = 2 \cdot \tau + n_{io}. \tag{3.3}$$

τ is the temporal lag, for which the autocorrelation function of $y(t)$ is small. n_{io} is the number of independent observations used to estimate the model error. In general n_{io} can be chosen to be larger than one, then the leave-one-out cross-validation becomes moving-block cross-validation (*Kunsch*, 1989). In this study we choose $n_{io} = 1$. In each cross-validation step (cv) $n_T = n - n_{lo}$ (T for training) data pairs $\{y_T, x_T\} := \{y\big(t_T(cv)\big), x\big(t_T(cv)\big)\}$ are used to estimate the parameters of the simple linear model. Thus in equation 3.2 it applies

$$a = a(y_T, x_T) \qquad \sigma_t(y)/\sigma_t(x) = \sigma_t(y_T)/\sigma_t(x_T) \qquad \bar{y} = \overline{y_T} \qquad \bar{x} = \overline{x_T}. \tag{3.4}$$

$\{y_V, \hat{y}_V\} := \{y(t_V(cv)), \hat{y}(t_V(cv))\}$ (V for validation) are then used to estimate the model error ϵ (equation 3.1). y_V is the central of the withheld observations in each cross-validation step cv and is considered as independent from the calibration process.

When the cross-validation process is completed, the skill score (SS) can be calculated as follows (e.g. *Wilks*, 2006):

$$SS = 1 - \frac{mse}{mse_r} \tag{3.5}$$

and (with equation 3.1)

$$mse = \frac{1}{n_{cv}} \cdot \sum \epsilon^2(cv) \qquad \epsilon(cv) = y_V - \hat{y}_V \qquad cv = 1, ..., n_{cv}. \tag{3.6}$$

mse_r is the mean of squared errors of the reference model, $\hat{y}_r$, as follows:

$$mse_r = \frac{1}{n_{cv}} \cdot \sum (y_V - \hat{y}_r)^2 \qquad y_r := \overline{y_T}. \tag{3.7}$$

SS, consisting of a contribution due to the correlation between the forecasts and observations, and two penalty terms relating to the reliability and bias of the forecast (*Murphy*, 1988), is considered as the more accurate skill measure than the correlation coefficient r^2 (*Wilks*, 2006). SS as specifically presented here can be considered as r^2 inflated by the conditional bias term (the unconditional bias is by construction constrained to be zero in least-squares regression; definitions of reliability and bias terms are given in the work of *Murphy*, 1988).

Finally, let $\overline{\cdot}^{cv}$ be the mean, and $\sigma_{cv}(\cdot)$ be the standard deviation of a variable over all cv repetitions, then the final model $\hat{y}_F$, with model uncertainty estimated by cross-validation (not to be mistaken for the model error $\epsilon(t)$) is

$$\hat{y}_F(t) = \overline{\hat{y}(t)}^{cv} \pm \sigma_{cv}(\hat{y}(t)). \tag{3.8}$$

Equations 3.1 to 3.7 apply similarly in multiple regression, where x has multiple columns (i.e., $x \in R^{(n \times p)}$). Note that, even though not shown in this study, SS as defined above is a powerful goodness-of-fit estimate especially in the case of multiple predictors, because it detects over-fitting (then $SS \leq 0$).

3.3.4 Example

Figure 3.4 provides a comprehensible example of the skill estimation procedure described in the previous section. The two plots show daily air temperature time series (y in equation 3.2) of the months January (top) and July (bottom) at AWS1 (blue line with stars). The red solid line is the linear model $\hat{y}$ as defined by equation 3.2. The example shows the model building and error estimation in an individual cross-validation repetition cv (equations 3.6 to 3.7). The grey bar indicates the n_{lo} observations left out in the model calibration (note that each cross-validation round cv, the grey bar is shifted one observation to the right). The amount of observations left apart is determined by the cross-validation parameter τ, which is the temporal autocorrelation in the time series. The values of τ are 9 in January, and 2 in July. This means that e.g. in the January time series an observation is considered independent from an other observation only if there is a shift of at least 9 time steps (in this case daily) between the two observations. Consequently the error ϵ_{cv} for each repetition cv (equation 3.6) is estimated as the difference between the central, independent observation (y_V, the blue star in the grey bar in figure 3.4) and the model value at this time step ($\hat{y}_V$, the red star in the grey bar in figure 3.4). y_r in equation 3.7 is the black star in figure 3.4 calculated as the mean of y_T, the observations used in the model training. Cross-validation is repeated until each observation is used once as y_V to determine the model error. This way independence between the observations used in the model training and the validation process is warranted, and at the same time all observations can be used to determine the final model (equation 3.8).

3.4 ESD model application

3.4.1 Downscaling parameters

In this section the above presented ESD procedure is applied to the daily AWS air temperature series in the Cordillera Blanca (introduced in section 3.2). As mentioned in section 3.3.3, the linear models are calibrated and validated independently for each calendar month and for each of the AWSs' time series. Thus the time series finally input to the ESD procedure (e. g., January at AWS1) consist of five months (consecutive Januaries) of daily data from the five years of available observations (July 2006 to July 2010), consequently approximately 150 observations per month and AWS (data gaps included) to calibrate the ESD model (e.g., the AWS1 Januaries's time series consists of 124 values and the AWS1 Julies's time series of 138 values, data

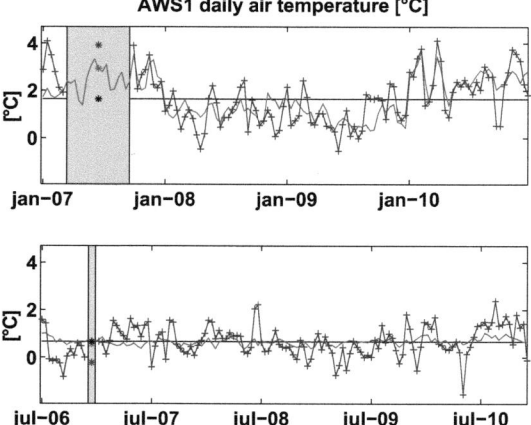

Figure 3.4: Example of skill estimation by cross-validation as described in section 3.3.3 for AWS1 predictands in January (top) and July (bottom). $y(t)$ (crosses-line), $\hat{y}(t)$ (solid grey), y_r (solid black), $y_V(cv)$ (star), $\hat{y}_V(cv)$ (red star), and $y_r(cv)$ (black star).

gaps excluded, in figure 3.4).

We define the standardization parameter R_σ as the ratio between the standard deviation of the predictand $\sigma_t(y)$ and the standard deviation of the predictor $\sigma_t(x)$

$$R_\sigma := \frac{\sigma_t(y)}{\sigma_t(x)} \tag{3.9}$$

R_σ is a parameter in the downscaling equation 3.2. Figure 3.5 shows the regression parameter α and the standardization parameter R_σ with uncertainties, estimated by cross-validation (equations 3.4 and 3.8), for each calendar month, and data from AWS1. Most remarkably both α and R_σ show high inter-monthly variability. Note that values of α close to one imply high covariance between the predictor and the predictand and values of α close to zero imply low covariance, because least-squares regression is applied to standardized variables (equation 3.1). Values of α are high for the months January to April (wet season in the Cordillera Blanca), and coincide with larger coefficient uncertainties than in the months with low values of α.

R_σ throughout the year varies from approximately 1.1 to 1.8. In December, the variability in the predictand data $\sigma(y)$ is almost twice the variability in the predictor data $\sigma(x)$, whereas in February the variability of the predictand is only 10% higher than the predictor variability. The variability of air500600 only slightly underesti-

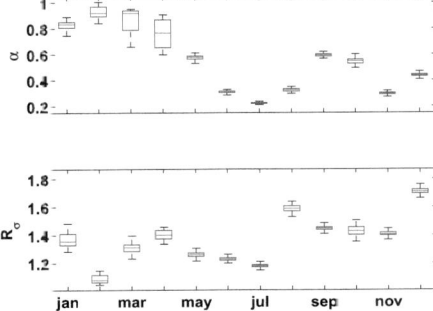

Figure 3.5: Medians (lines in within the boxes) of the downscaling model parameters α (top) and R_σ (bottom) estimated by cross-validation for the AWS1 predictands and all calendar months. The edges of the blue boxes are the 25th and the 75th percentiles. The black dashes (outside the boxes) extend to the most extreme data not considered as outliers.

mates the measured air temperature variability also in the dryer months May, June and July, even if in those months low values of α indicate a weak relationship to the predictor. The high inter-monthly variations in the downscaling parameters clearly support the importance of using different models for the different calendar months.

Values of the cross-validation parameter τ are shown in figure 3.6 for all months and AWSs. As defined in section 3.3.3, τ can be interpreted as temporal autocorrelation, or persistence (*Von Storch and Zwiers*, 2001; *Madden*, 1979) (here in days). τ varies from 1 (no persistence) to 20 (high persistence) days, primarily as a function of calendar month, but also within the same month for the different AWSs. At all AWSs, τ has generally lower values in the second half of the year, the drier months June to December ($1 \leq \tau \leq 5$), and higher values from January to May ($3 \leq \tau \leq 20$). Higher values of τ especially in the wet months are probably related to rainy episodes; i. e. sequences of rainy days followed by sequences of dry days, where wet and dry sequences have similar durations. Typical lengths of such episodes in the tropics (i.e., the number of wet and dry days) range from 30 to 60 days (equivalent to $2 \cdot \tau$), with the basic mechanism known as the Madden-Julian Oscillation (MJO) (*Madden and Julian*, 1994). The MJO develops generally strongest during austral summer (wet season in the Cordillera Blanca). This represents a possible explanation why the values of τ are higher in the wet season than in the dry season. For the Bolivian Altiplano (located nearby the Cordillera Blanca), *Garreaud et al.* (2003) report synoptic periods of 15 days. Note that small values of τ in figure 3.6 indicate not only small inter-daily (intra-seasonal), but also small inter-annual variability. Consequently there are no important differences amongst the different years of the respective calendar months. In fact ENSO, the most important source of inter-annual variability in the region, has its strongest and most widespread impacts during austral summer.

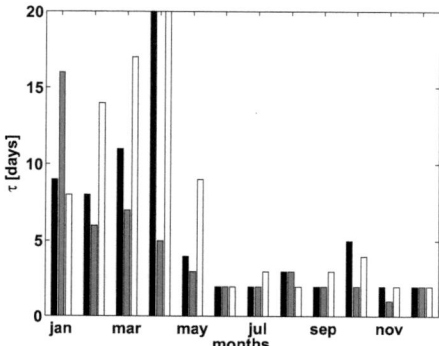

Figure 3.6: Decorrelation time τ of daily air temperature observations at AWS1 (black), AWS2 (grey), and AWS3 (white) for each month of the year.

Regarding the differences of τ amongst the different AWSs, generally the AWS1 and the AWS3 time series show rather similar values, whereas AWS2 shows generally smaller values of τ. A possible explanation is that, even if AWS2 is close to AWS1, AWS1 and AWS3 are located in more similar altitudes, rather exposed to the free atmosphere. AWS2 is located approximately 200 m lower than AWS1 and AWS3, close to the glacier and thus more influenced by the glacier boundary layer. However, the large differences in τ in April between AWS2 ($\tau = 5$) and the other to AWSs ($\tau = 20$), are possibly related to erroneous measurements at AWS2. The site is less exposed, located close to the glacier basin, with lower wind speeds than at AWS1 and AWS3. Even if the air temperature sensors are all ventilated, overheating is more likely to occur at a site with low wind speeds. Finally, if due to overheating larger diurnal cycles are measured, the intra-seasonal or inter-annual variations become less important and consequently there are lower values of τ.

3.4.2 Skill assessment

Figure 3.7 shows values of SS for the three AWSs' air temperature time series and each calendar month. Similar to the downscaling parameters, for all AWSs' time series a similar, distinct seasonal pattern of ESD model skill is evident. Values of SS are high from January to May and September to October, and low in June to August and November to December. E.g., at AWS1 the difference between the minimum and the maximum is from almost no skill ($SS \simeq 0.1$) to relatively high skill ($SS \simeq 0.7$). The highest values of SS for all AWSs are in February (the core wet season in the Cordillera Blanca), and the lowest values in July (dry season).

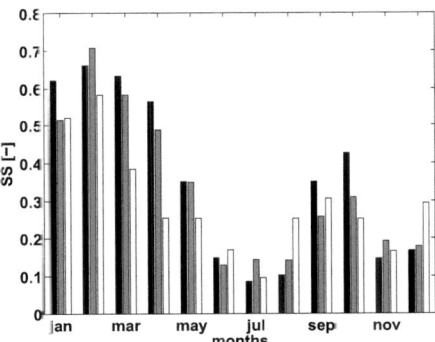

Figure 3.7: Values of SS for AWS1 (black), AWS2 (gray), and AWS3 (white) for the predictand $air500600$ estimated by cross-validation (as described in 3.3.3).

Thus in the wet season, the largest portion of inter-daily variability in the AWS air temperature time series can be explained by the single large-scale NCEP predictor $air500600$. These high values of SS at a diurnal time scale are even more remarkable if we consider that the reanalysis data predictor corresponds to a horizontal area of approximately $4 \cdot 10^4 km^2$ (the NCEP reanalysis grid point distance is approximately 200 km) extending over two vertical levels, and that the Cordillera Blanca is completely misrepresented in the NCEP model topography. However for the Bolivian Altiplano, *Garreaud et al.* (2003) similarly report spatially very coherent intra-seasonal weather patterns (superimposed to a persistent east-west moisture gradient). In fact, the MJO is known to act on large spatial scales, with local wavelengths of $1.5 - 2 \cdot 10^3 km$.

By contrast in the dry season, the reanalysis data predictor $air500600$ shows almost no covariance to the local data. The analysis in section 3.4.1 indicates that intra-seasonal and inter-annual variations play a minor role in these respective months. We conclude that variability must be governed by local processes, triggered by the strong radiation interacting with the complex topography, in a way that the generally weaker synoptic forcing at this time of the year has almost no impacts. Figure 3.4 shows that for the dry month July the model chosen by least-squares regression is almost constant, as there is no co-variability amongst the predictor and the predictand data series. To summarize, in the dry season the observed inter-daily temperature variations show no link to large-scale air temperature variations, whereas in the wet season, where generally larger variations occur (figure 3.3), the large-scale predictors explain even the largest portion of the local-scale variations.

Values of SS and r^2, averaged over all months for the different AWSs, are displayed in figure 3.8. On average SS is highest at AWS1 and lowest at AWS3, but for August and December, SS at AWS3 is more than 10% higher than for AWS1 and AWS2 (figures 3.7 and 3.8). Figure 3.8 shows that here values of r^2 only slightly overestimate SS (note however that the differences become considerably larger in the case of multiple predictors).

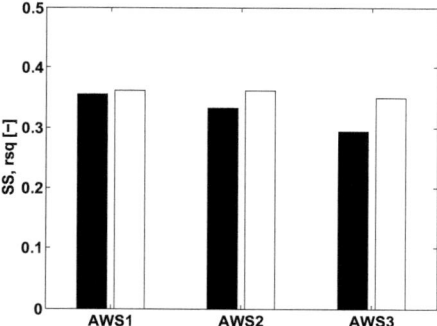

Figure 3.8: Averaged values of SS (black) and r^2 (white) over all months of the year for AWS1, AWS2, and AWS3 (from left to right).

3.4.3 Results for different time scales

In this study the short length of the data represents a lower limit of possible time resolutions for the skill assessment, because for lower temporal resolutions the number of observations n available for the model set-up decreases. More specifically, the five-years observations in this study include a time series of 50 monthly means, or 1575 daily means. As we use separate models for each month, the time series include only five monthly means, but approximately 150 values in the daily time scale. With this regard it makes sense to profit from the higher (daily) temporal resolution to have more observations for the model fit. In practice however, we question if there is daily information in the reanalysis data relevant at the scale of the observations, or if the optimum skill of reanalysis data (with regard to local-scale variability) is found at lower time resolutions.

In this section we repeat the modeling procedure as introduced in section 3.3.3, but for different temporal resolutions. The number of observations n in each time scale is set constant. The lowest temporal resolution where the available month-separated AWS time series still include enough values for the model calibration/cross-validation are 5-daily means ($n = 13$).

Figure 3.9 shows values of SS for decreasing time scales (from daily means to 5-daily means), and the different calendar months on the example of AWS1 data. At each time scale SS averaged over the twelve calendar months is also shown. The results show that with decreasing time scale values of SS generally become larger. However there is no clear linear increase evident for the individual months. The means of SS over all months facilitates the interpretation of the twelve different calendar months' cases, showing a general increase with decreasing time resolution. The largest mean increase of SS is evident from the 3-days to the 4-days time steps. To sum up, this analysis suggests that predictability based on large-scale predictors increases with decreasing time resolution. Consequently, when longer time series are available we can suggest to use longer averaging windows in order to obtain higher skill.

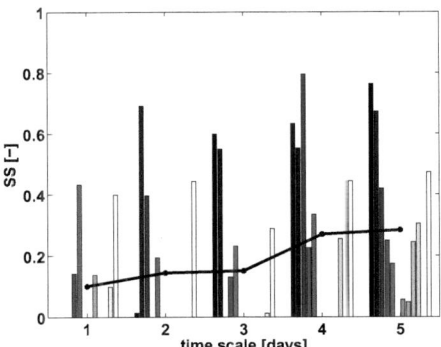

Figure 3.9: SSs for different months (from left to right within each column, or black to white, respectively: January to December) and different time scales (from 1-day averages to 5-day averages). Note that, because here $n = 13$, versus $n > 100$ as in the previous sections, the values of SS for 1-daily means do not correspond to the results in figure 3.7. The black points connected by the bold line shows SS averaged over all months. The values are shown for the AWS1 data series. Negative values of SS are not shown.

3.4.4 Towards automated predictor selection

To this point we presented the ESD model with a predictor selected by arguments independent of the data (a priori predictor selection). In this section we compare the performance of the a priori selected predictor to a list of other potential predictors, to show how the skill assessment presented here can be used for data-based predictor selection. We examine the relative performance of the following predictor variables: air temperature at the NCEP model surface ($airSFC$), sea level pressure (slp), geopotential height at the 600 hPa level ($gph600$), air temperature at the 600 hPa level ($air600$), air temperature at the 500 hPa level ($air500$), air temperature averaged over the 18 closest grid points in the 500 and 600 hPa levels ($air18gp$), and the a priory selected predictor $air500600$. All predictors, except $air500600$ and $air18gp$, are an average of the four closest grid points (listed in table 3.1) at the respective pressure levels. slp and geopotential height are frequently applied predictors in ESD. We use $air500$ and $air600$ to assess whether the four-grid point predictors fit better to the data then the eight- or eighteen-grid point predictors $air500600$ and $air18gp$, to obtain a first-order, data-based estimate for the optimum spatial scale in this case.

Figure 3.10 shows values of SS for AWS1 air temperature time series for each calendar month, and all the above mentioned predictors. The two predictors $air18gp$ and $air500600$ show the highest values of SS in almost all months. Even if the differences in SS to the predictors $air600$ and $air500$ are not very large, the results suggest that the forecast skill can indeed be improved by using coarser-scale information (eight- or sixteen-grid point average instead of four-grid point average). The exact amount of grid points in this study can be seen as reference for the skillful scale of NCEP reanalysis data, however we expect it to largely vary for different GCMs (*Benestad et al.*, 2008). Averaged over all months, $air500600$ shows a marginally higher value of SS than $air18gp$. $air500$ shows generally higher values of SS than $air600$. $airSFC$ clearly shows lower values of SS than $air600$, $air500$, or $air500600$ in all months. Thus the NCEP reanalysis variables that are less affected by the model surface actually emerge as the better predictors, even for surface variables. This supports our a priori assumption that unrealistic surface information negatively affects the data. The two circulation predictors slp and gph show almost no skill in all months ($SS \leq 0.1$). We have also assessed predictors such as the horizontal wind velocities u and v, and specific humidity sh at the 600 hPa levels (analysis not shown here), because they are variables important for the weather and climate of the tropical Andes (*Garreaud et al.*, 2003, 2009). These variables have shown only very low skill as single predictors

($SS < 0.1$). Note however that, even if these variables show no skill as single linear predictors, they can have significant importance in more complex ESD models (*Hofer et al.*, 2010).

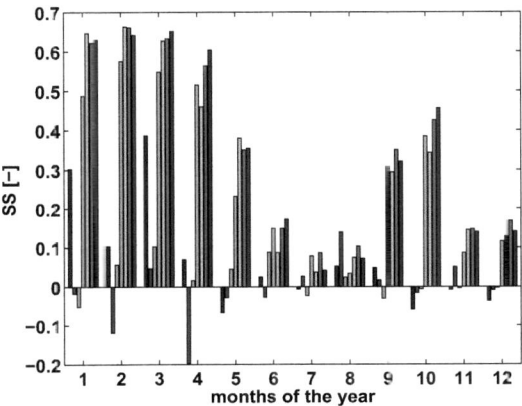

Figure 3.10: Values of SS for each month (here for AWS1), and different predictors (from left to right): $airSFC$, slp, $hgt600$ (light blue), $air600$ (green), $air500$ (yellow), $air500600$, and $air18gp$.

We conclude from figure 3.10 that (1) the a priori predictor choice is clearly supported by the data, (2) even if values of SS show a distinct seasonal cycle, the same predictors show the highest skill throughout the year, (3) spatial averages of several grid points show increased skill as predictors compared to fewer-grid point averages, (4) circulation predictors such as slp and gph show almost no covariance with AWS air temperature at a diurnal time scale.

3.4.5 Accounting for the effects of diurnal periodicity

Finally we show a simple analysis based on the AWS1 air temperature data to demonstrate the effects of periodicity (in this case diurnal) for regression analysis based solely on r^2 (an often applied criterium for predictor selection). Figure 3.11 shows values of r^2 between 6-hourly, (all-month) time series of AWS1 air temperature, and the predictors assessed in section 3.4.4: $airSFC$, slp, $hgt600$ $air600$, $air500$, $air600500$, and $air18gp$ Furthermore the same analysis is shown at a daily time scale (thus r^2 between the same time series, but the 6-hourly data averaged to daily means). Results show that in the 6-hourly data (left hand side in figure 3.11) $airSFC$

clearly appears as the best predictor, showing relatively high covariance ($r^2 > 0.5$), whereas all other predictors show only small covariance ($r^2 < 0.2$) with the sub-daily predictand time series. This pattern significantly changes in the daily time scale (right hand side in figure 3.11): now $air500600$ shows the highest covariances ($r^2 = 0.4$), and accordingly air $18gp$, $air500$ and $air600$ appear as important predictors, but $airSFC$ shows almost no covariance ($r^2 = 0.1$), similar to slp (but still higher than $hgt600$ with $r^2 < 0.05$ at both time scales). In the diurnal analysis $airSFC$ appears as important predictor only because it has the most pronounced diurnal variations which explain the largest portion of variability also in the predictand data (i.e., high values during day-time and low values in the night). However this is achieved easily with a constant diurnal cycle. E.g., the covariance between the hourly air temperature series and a time series composed by consecutive constant diurnal cycles is $r^2 = 0.7$. For these reasons predictor selection that does not account for diurnal (or other) periodicity is not meaningful. This analysis is by no means innovative in statistics, but the issue of periodicity is not accounted for appropriately in numerous studies.

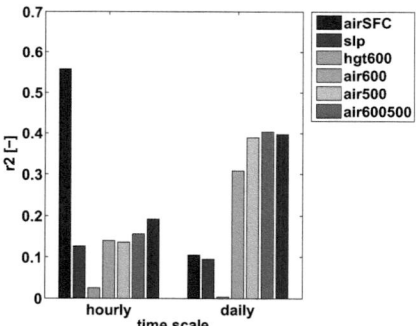

Figure 3.11: r^2 for 6-hourly and daily (all-month) time series of AWS1 air temperature and the predictors (from left to right): $airSFC$, slp, $hgt600$, $air600$, $air500$, $air600500$, and $air18gp$.

3.5 Summary and conclusions

We have presented a MOS technique that links large-scale data predictors to Gaussian target variables. The method is appropriate for temporally high-resolution time series (i.e. sub-daily to daily time scale), as it accounts for persistence in the data series. We highlight the importance of systematically eliminating seasonal (or diurnal, respectively) periodicity in meteorological time series, and on the same time in the ESD model error, by using separate models for the different times of year (and times of day). The presented skill estimation based on cross-validation allows for the usage of each observation both in the calibration and in the validation process and is therefore useful especially when only short-term (i.e. few-years) observational time series are available.

We have shown an application of the ESD model with NCEP reanalysis data predictors and daily air temperature predictands measured at high-altitude sites in the glaciated Cordillera Blanca (Peru). High seasonality of statistical data properties (e.g. persistence), and ESD model parameters, emphasizes the importance of using different models for different times of year. The ESD model skill shows high seasonality as well, with generally large skill in the wet season (January to March) and little to almost no skill in the dry season (June to August). We conclude that in austral summer the generally stronger synoptic forcing leads to high predictability of local-scale variables based on large-scale predictors. By contrast, during austral winter local-scale meteorology is governed by radiation-induced, dynamic processes, where local topography and surface properties (not present in the large-scale model) play an important role.

We have defined the a priori (non data-based) predictor air temperature averaged over eight grid points around the study site ($air500600$). The a priori predictor clearly shows higher skill than other potential predictors, such as $airSFC$ or slp. We conclude that large-scale surface variables are weak predictors if the model surface is not representative for the real surface, because unrealistic boundary layer variability masks the relevant synoptic forcing. The eight-grid point predictor shows higher skill than the same predictor averaged over four grid points. This indicates that spatial averaging of large-scale predictors increases the ESD model skill as it reduces errors related to single grid point data. However, the optimum scale of spatial averaging is expected to largely differ amongst different models.

In a further experiment we have decreased the ESD time resolution stepwise from 1-day to 5-day averages. The results show increasing ESD model skill for decreasing time resolutions. Consequently we suggest switching to lower temporal resolutions when the ESD model skill is low, given that long enough data series are available.

The presented ESD model can be generalized to non-linear, multiple regression problems (not shown here). The validation process is especially useful in multiple predictor fitting because it detects over-fitting. The method is not restricted to reanalysis data and can be applied to any large-scale data predictors. However, we recommend the application as MOS technique (i.e., using the same data source in the calibration process and the forecast) since otherwise the validation results do not account appropriately for potential model errors.

CHAPTER IV

Paper 3: Comparing the skill of different reanalyses as predictors for daily air temperature on a glaciated mountain (Peru)

Marlis Hofer, Ben Marzeion, and Thomas Mölg

Faculty of Geo- and Atmospheric Sciences, University of Innsbruck (Austria)

(at the time of printing, since 06/2011 under review by *Climate Dynamics*)

4.1 Introduction

Even though reanalysis is the most accurate way to interpolate atmospheric data in time and space (i.e., using the methods of numerical weather prediction), its usefulness to document climatic trends and variability is debated (e.g., *Bengtsson et al.*, 2004; *Kalnay et al.*, 1996). Uncertainties in reanalysis data derive from errors in the atmospheric models used to generate the background forecast (or first guess) for the data assimilation. Also, the observations used in the assimilation process to adjust the model solution towards the true atmospheric state are incomplete in time and space, and have errors as well. Especially changes in the observation system (e.g., the introduction of satellite data in the late 1970s; or changes of observation density) cause artificial climate variability and trends and thus restrict the usefulness of reanalysis data in some climate studies (*Trenberth et al.*, 2001; *Bengtsson et al.*, 2004). Furthermore, reanalysis data (as assimilated data in general) are not physically consistent, which limits their usability e.g. in constituent transport and hydro-meteorological cycle studies (e.g., *Bey et al.*, 2001). Reanalysis data documentation articles and

Table 4.1: Summary of available reanalyses by different institutions.

	NCAR	ERA40	Interim	JCDAS	MERRA
Generation	1st	2st	2st	2st	2st
Status	operated	completed	operated	operated	operated
Period	1948-	1957-2002	1979-	1979-	1979-
Spatial res.	T62 L28	T159 L60	T255 L60	T106 L40	$2/3 \times 1/2$ L72
Temporal res.	6-hourly	6-hourly	6-hourly	6-hourly	3-hourly
System	3D-Var	3D-Var	4D-Var	3D-Var	4D-Var
Institution	NCEP	ECMWF	ECMWF	JMA	NASA

many other studies report about these limitations (e.g., *Kalnay et al.*, 1996; *Rood and Bosilovich*, 2009; *Trenberth et al.*, 2001; *Uppala et al.*, 2005).

Global reanalysis data are generated at four institutions worldwide: NCEP (National Centers for Environmental Prediction), ECMWF (European Centre for Medium-Range Weather Forecasts), JMA (Japan Meteorological Agency), and NASA (National Aeronautics and Space Administration), in cooperation with partner institutions not mentioned here for brevity. An overview of the most important available reanalyses is given in table 4.1. Second-generation reanalyses have profited from advances in computing power and modeling systems, from better treatment of the assimilated observations, and other lessons learned from problems in the earlier projects (e.g., *Rood and Bosilovich*, 2009). Moreover, with increasing computer power available, higher performance 4D-Var (4-dimensional variation analysis) systems became feasible for reanalysis, replacing 3D-Var (3-dimensional variational analysis). Spatial resolutions of global reanalysis span from triangular truncations T62 to T255 and higher, with 28 to 72 levels in the vertical (cf. table 4.1). Temporal resolutions are 6-hourly or higher for all reanalyses. Given the problems caused by major changes in the observing systems, more recent reanalyses are restricted to data-rich periods in the satellite era (e.g., 1979 onwards), but some reanalyses also exist that include pre-satellite periods from 1948 onwards. Each institution provides one reanalysis stream operationally (i.e., with availability up to present).

Studies exist that compare different reanalysis data in some regards. *Simmons and Jones* (2004) evaluate trends and low-frequency variability in surface air temperature of ERA-40 (the 45 years ECMWF reanalysis) and NNRP (the NCEP/NCAR reanalysis) with CRU (Climate Research Unit) (*Jones and Moberg*, 2003) data sets globally. *Dessler and Davis* (2010) analyze NNRP, ERA-40, JRA-25 (the Japanese

25-yr reanalysis), MERRA (the Modern Era Retrospective-Analysis for Research and Applications from NASA), and ERA-int (the ECMWF-interim reanalysis) with regards to tropospheric humidity trends. They find artificial negative long-term trends in NNRP tropospheric humidity and large bias in NNRP tropical upper tropospheric humidity not evident in all the other reanalyses. *Bosilovich et al.* (2008) show that reanalysis precipitation improves in recent systems and that ERA-40 products show reasonable skill over Northern Hemisphere continents, but less so in the tropical oceans, whereas JRA-25 shows good agreements in both tropical oceans and Northern hemisphere continents. *Trenberth et al.* (2001) study the quality of ERA-15 (the 15-yrs ECMWF reanalyses) and NNRP air temperatures in the tropics, finding that ERA-15 show large discrepancies to observations due to changes in the satellite system, whereas NNRP show good agreement. All studies report about important differences amongst different reanalysis types.

In this study we compare the skill of different reanalyses as predictors for site-specific, daily air temperature in the tropical Cordillera Blanca (cf. figure 1.1). The Cordillera Blanca is a glaciated mountain range in the Northern Andes of Peru, harboring 25% of all tropical glaciers with respect to surface area (e.g., *Kaser and Osmaston*, 2002). The glaciers have heavily shaped the socioeconomic development in the extensively populated Rio Santa valley, with the occurrence of several disastrous glacial lake outburst floods and ice avalanches (e.g., *Carey*, 2005, 2010). Beyond, melt water from the currently shrinking glaciers (e.g., *Ames*, 1998; *Georges*, 2004; *Silverio and Jaquet*, 2005) is an important water source for agriculture, households and industry in the dry season (e.g., *Mark and Seltzer*, 2003; *Kaser et al.*, 2003; *Juen*, 2006; *Juen et al.*, 2007; *Kaser et al.*, 2010), when precipitation is extremely scarce (*Niedertscheider*, 1990).

To quantify the impacts of future climate change to glaciers in the Cordillera Blanca is thus of primary relevance for the population. Due to the absence of long-term, high-resolution atmospheric measurements in the Cordillera Blanca, however, longer-term, process-based assessments of the glacier-atmosphere link (*Mölg et al.*, 2009) is problematic. *Hofer et al.* (2010, 2011), by means of empirical-statistical downscaling, explore the potential of NNRP to provide more knowledge about past atmospheric variations in the Cordillera Blanca with promising results. In the present study it is the goal to identify the most appropriate data set, beyond NNRP out of all available reanalyses, for the study site and target variable.

Whereas we do not claim the results being valid outside the study area or for different variables, the method presented here provides the basis for further inter-comparison studies of reanalysis data, and large-scale model output in general, against point-observations. In section 4.2 we present the data sets used in this study. Section 4.3 provides an overview of the applied methodology. Finally, we show the results in section 4.4 and the conclusions in section 4.5.

4.2 Data

The skill assessment of reanalyses in this study is based on local air temperature measurements carried out at a high-altitude site in the glaciated Cordillera Blanca mountain range (figure 1.1). In earlier studies, we have focused on skill assessment of NNRP, using air temperature and specific humidity measurements from multiple sites in the Cordillera Blanca (*Hofer et al.*, 2010, 2011). In *Hofer et al.* (2011), we report about the spatial homogeneity of air temperature data measured throughout the Cordillera Blanca. We find that differences in NNRP skill with regard to different automated weather stations (AWSs) are small, with the same seasonality of skill being evident for all AWSs. In this study, it is therefore reasonable to use solely one, but the longest and most reliable, quality-controlled air temperature time series from all AWSs in the Cordillera Blanca (described in more detail by *Juen*, 2006; *Hofer et al.*, 2011), hereafter referred to as airt-CB. In *Hofer et al.* (2011) we find, for an air temperature time series presumably affected by radiation errors, lower skill of NNRP than for the more reliable observational time series and conclude that measurement errors negatively affect the skill assessment (i.e., underestimation of skill).

The AWS providing airt-CB is located on a moraine at 5000 m a.s.l. in the Paron valley (Northern Cordillera Blanca, cf. figure 1.1). airt-CB is measured with a HMP45 sensor by Väisalla in a ventilated radiation shield (described by *Georges*, 2002). airt-CB is available from 07/2006 to 07/2010. In figure 3.1, two further sites are indicated, where AWSs exist on and next to glaciers: Paria, located close to the Paron valley but East of the main divide, and Shallap in the Southern Cordillera Blanca, West of the main divide (*Juen*, 2006; *Hofer et al.*, 2011).

In this study we consider four different reanalyses: (1) NNRP, (2) ERA-int, (3) JCDAS (the JMA Climate Data Assimilation System reanalyses), and (4) MERRA (see table 4.1 for details about the data sets). These are the only available reanalysis data that cover the period of available measurements in this study, provided operationally at the respective institutions NCEP, ECMWF, JMA and NASA. All data are downloaded on their native spatial grids, in an area extending from 5°N to 20°S and 90°W to 65°W, and in the 400 to 700 hPa levels.

4.3 The method

We apply the method of skill assessment presented in detail in *Hofer et al.* (2011). The procedure first includes the calibration of simple linear regression models ($\hat{y}$) between large-scale model predictors (x), and observations, or target variables (y),

$$y_s(t_m) = a_m \cdot x_s(t_m) + \epsilon(t_m), \tag{4.1}$$

where index s denotes the variables being standardized, m indexes the calendar month ($m = 1, ..., 12$), and ϵ is the model error, obtained as the difference between y and $\hat{y}$

$$\epsilon(t_m) = y_s(t_m) - \hat{y}_s(t_m). \tag{4.2}$$

From equations 4.1 and 4.2 it is apparent that

$$\hat{y}_s = a_m \cdot x_s(t_m). \tag{4.3}$$

Note that the regression parameters, a_m, differ for different calendar months. To use individual models for each month is important in order to account for seasonality both in the daily time series, and in the model error.

The skill estimation is then based on a modification of leave-one-out cross-validation that accounts for autocorrelation in the daily time series and therefore guarantees complete independence between training and test data. This is an effective validation procedure especially designed for short-term and high-resolution time series. The skill score, SS (*Wilks*, 2006), can be calculated as

$$SS = 1 - \frac{mse}{mse_r} \tag{4.4}$$

based on mse, the mean squared error

$$mse = \frac{1}{n_{cv}} \cdot \sum \epsilon_{cv}^2 \qquad \epsilon_{cv} = y_{s,V} - \hat{y}_{s,V}(x_{s,V}) \tag{4.5}$$

and mse_r, the mean squared error of the reference model, y_r, as follows

$$mse_r = \frac{1}{n_{cv}} \cdot \sum (y_{s,V} - \hat{y}_r)^2 \qquad \hat{y}_r := \overline{y_{s,T}}. \tag{4.6}$$

Above, ϵ_{cv} is the difference between independent observation $y_{s,V}$ and model, $\hat{y}_{s,V}(x_{s,V})$ (V for validation), obtained for the cross-validation repetition cv, with

$cv = 1, ..., n_{cv}$ (n_{cv} is the number of cross-validation repetitions). $\hat{y}_r$ is defined as the mean of all observations used for the model training, $y_{s,T}$ (T for training).

Skill estimation is repeated for $\hat{y}_s(x)$ based on predictors x from all four reanalysis data assessed in this study, NNRP, ERA-int, JCDAS, and MERRA. Evaluating $\hat{y}_s(x)$, as defined above, rather than the untransformed predictors, x, has the following important implications. (1) The skill assessment is focused on performance of the reanalysis predictors in capturing intra-seasonal, and inter-annual variations, rather than the seasonal cycle. This is important because seasonal variations are generally larger than inter-annual and intra-seasonal variability and would otherwise dominate the results. (2) The skill assessment does not penalize for differences in monthly means and variances between reanalyses and observations. This allows for more general inter-comparisons of predictor variables from different levels (performed in this study), or with different physical units (performed in *Hofer et al.*, 2011).

Regarding the choice of predictor variables, we select air temperature as the priori predictor for local-scale air temperature. By contrast to data-based selections, a priori selections are based on information outside the data (i.e. prior to data analysis) and therefore provide the appropriate basis for data-based model inter-comparison studies (*Hofer et al.*, 2011).

In terms of the optimum downscaling domain, *Hofer et al.* (2011) find for NNRP that averaging over eight grid points surrounding the study site (i. e., four grid points in two vertical levels) increases the skill relative to using four-point averages from a single level, or more than eight grid points. For NNRP the optimum averaging domain extends 2.5×2.5° horizontally over the two vertical levels 500 and 600 hPa.

Grid point averaging of atmospheric models to obtain higher skill predictors can be considered as a compromise between minimizing numerical model errors related to single grid point data (*Grotch and MacCracken*, 1991; *Willamson and Laprise*, 2000; *Räisänen and Ylhäisi*, 2011) and loosing climate information at the minimum model scale (i. e., the distance of two neighboring grid points). Note that we suspect the latter effect being less dominant for the predictand air temperature in the Cordillera Blanca, than it might be for other sites, due to the pronounced spatial homogeneity of synoptic forcing in the region (e.g., *Garreaud et al.*, 2003; *Hofer et al.*, 2011).

Hofer et al. (2011), however, assess only four different spatial domains in their study, and the optimum domain for NNRP not necessarily applies for the other reanalyses. In this study we therefore conduct a systematic assessment of model skill as a function of spatial averaging, for all four reanalyses. We seek to identify the optimum spatial domain for which each of the four reanalyses shows the highest skill for local-scale air temperature predictands. After determining the optimum scale for each reanalysis, their performances relative to each other are compared at the individual optimum scale.

The optimum scale analysis, where we distinguish between horizontal and vertical domain extensions, is done as follows. For each reanalysis, the grid point located closest to the study site is identified and the skill assessment is conducted for the single grid point predictor, as explained above. Thereafter, the horizontal domain of averaging is increased consecutively by the minimum scale of each reanalysis (the minimum scale is 2.5° for NNRP, 0.72° for ERA-int, 1.25° for JCDAS, and 0.5° for MERRA, cf. table 4.1) and the skill assessment is repeated for each domain. Then the two closest vertical levels are added, and the analysis is started over by first considering only the horizontally closest grid point (now in more levels) and then consecutively increasing the horizontal area (as it was done for the single-level domain before). Table 4.2 shows an overview of all horizontal domain/vertical level combinations considered in this study. Number of grid points, scales of the horizontal domains, as well as vertical levels, change for the different reanalyses because of their different spatial resolutions. For MERRA, in particular, the number of grid points n_{gp} for domain n is not like for the other reanalyses $n_{gp} = n^2$, because latitudinal and longitudinal grid resolutions of MERRA are not the same (1/2° versus 2/3°).

In the optimum domain analysis of this study we disregard the assessment of remote grid point predictors, as proposed e.g. by *Brinkmann* (2002). First, the selection of remote grid points as predictors is necessarily related to data-based selections, while for the sake of model inter-comparison in this study we restrict to a priori selections as far as possible. Second, we assume that for more skillful models, higher skill is found close to the study site. I.e., the closer a model approaches reality, the smaller the differences are between model and (unbiased) observations (e.g., the optimum scale becomes very small), and thus the closer the optimum grid point approaches to the real geographical location of the study site. However, in empirical-statistical downscaling studies that consider remote grid point information we recommend using principal component (PC) analysis (e.g., *Hannachi et al.*, 2008; *Schubert and Henderson-Sellers*, 1997; *Huth*, 2004; *Hofer et al.*, 2010), as PC analysis effectively separates important atmospheric modes from noise in multi-dimensional data sets.

Nevertheless we perform a data-based optimum scale analysis for all reanalyses because it gives important insight to the performance of the individual reanalyses (i.e., the larger the discrepancy between minimum scale and optimum scale, the larger are

Table 4.2: Vertical levels (hPa) and horizontal domains considered for each reanalysis. In the case of sl (single-level averages), all listed levels (e.g., a:b:c) are individually considered (b, the value between the two colons, is the vertical resolution). In the case of ml (multiple-level averages), averages over the levels in the list are considered. In terms of horizontal domain, the number in front of gp indicates the number of grid points of each averaging domain, the value in brackets indicates the scale of the respective domain. The horizontal domains are increased until a maximum scale of 25°. In our analysis, each of the horizontal domains is combined with each of the vertical level combinations.

	NCAR	Interim	JCDAS	MERRA
Vertical levels	sl-	sl-	sl-	sl-
	400:100:700	400:50:700	400:100:700	400:50:700
	ml-	ml-	ml-	ml-
	500:100:600	500:50:600	500:100:600	500:50:600
	400:100:700	450:50:650	400:100:700	450:50:650
		400:50:700		400:50:700
Horizontal domains	1gp(2.5°)	1gp(0.72°)	1gp(1.25°)	1gp(0.5°)
	4gp(5°)	4gp(1.44°)	4gp(2.5°)	2gp(1°×0.5°)
	9gp(7.5°)	9gp(2.16°)	9gp(4.75°)	9gp(1.5.°)
	16gp(10°)	16gp(2.88°)	16gp(6°)	12gp(2°×1.5.°)
	...	...	...	...

the errors related to numerical noise; and the larger the optimum scale in general, the lower the performance of the reanalysis system can be assumed).

4.4 Results

When interpreting values of SS, please note, it can be shown that SS consist of three terms

$$SS = r_{\hat{y}y}^2 - \left(r_{\hat{y}y} - \frac{\sigma_{\hat{y}}}{\sigma_y}\right) - \frac{\mu_y - \mu_{\hat{y}}}{\sigma_{\hat{y}}} \qquad (4.7)$$

where σ_y and $\sigma_{\hat{y}}$ are the standard deviations of the target variable y and the model $\hat{y}$, respectively, and μ_y, and $\mu_{\hat{y}}$ are the means. The first term on the righthand side of equation 4.7, r^2, is the correlation coefficient, the second term is a penalty due to errors in estimating the variance (the reliability of the forecast), and the third term is another penalty that accounts for the model bias (unconditional bias) (*Wilks*, 2006). SS is thus the more complete goodness-of-fit measure than r^2. In this study SS is estimated based on the test error from cross-validation.

4.4.1 Optimum scale analysis

Figure 4.1 shows values of mean SS (i.e., values of SS averaged over all calendar months) for increasing spatial domains in different vertical levels, or level combinations, for each of the predictors NNRP, ERA-int, JCDAS, and MERRA.

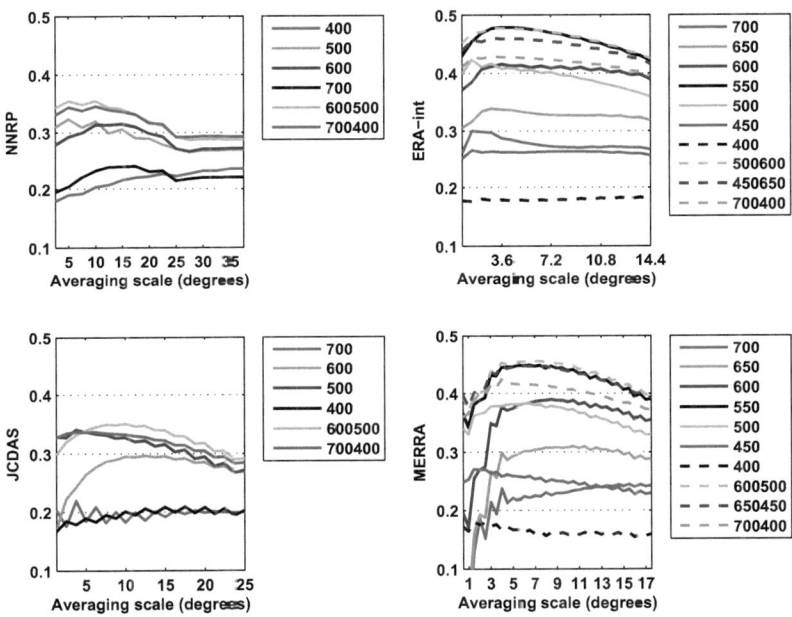

Figure 4.1: Values of SS averaged over all calendar months for different vertical levels and combinations (different colors and lines) and increasing horizontal domains (from left to right), for NNRP (top left), ERA-int (top right), JCDAS (bottom left), and MERRA (bottom right) predictors. Please note that the scale on the abscissa changes for the different reanalyses, because of the different grid point spacings (the scale on the ordinate is kept fix).

In the case of NNRP, values of mean SS with increasing domain size first increase, and then decrease. The highest mean skill is found at a scale of 5° (thus only $n_{gp} = 4$ grid points to be averaged). At a scale of 25°, mean SS does not further decrease, but remains more or less constant. The two single levels with the largest differences to the study site's vertical level show the overall lowest SS, whereas the two levels located close to the study site show comparably high SS. As found by *Hofer et al.*

(2011), by averaging the same horizontal domain over two or more levels, higher skill is achieved than by using single-level domains.

Also for ERA-int, levels located closer to the study site show higher values of SS, and the decrease in skill with increasing vertical distance from the study site is even larger than for NNRP. Averaging over the levels between 500 and 600 hPa increases the skill slightly for small scales, but overall, the 550 hPa level shows the highest skill. In terms of horizontal domain, the highest skill is found at a scale of 2.88° ($n_{gp} = 16$). Similar to NNRP, values of mean SS increase significantly with increasing domain until reaching the maximum SS and then decrease slowly. This indicates, most importantly, that by averaging too many grid points less information is lost, than by using too few grid points.

For JCDAS, with 8.75° ($n_{gp} = 49$), a considerably larger optimum scale results than for both NNRP and ERA-int, even if the minimum scale is smaller for JCDAS than for NNRP. The 500 to 600 hPa average shows the highest skill of all levels and the skill decreases with increasing vertical distance of the considered level from the study site, however not so much as for ERA-int. The increase in skill with increasing domain scale is more important than the decrease after reaching the optimum scale, as for the other reanalyses.

For MERRA, in terms of horizontal area, the highest skill is found at a scale of 7.3°. Because of the high horizontal resolution of MERRA this includes by far the largest amount of grid points to be averaged compared to the other reanalyses ($n_{gp} = 154$). Consequently, even if the spatial resolution of MERRA is the highest of all reanalyses, the optimum scale found for MERRA is even lower than for the coarse-resolution NNRP. This implies considerable uncertainty related to numerical noise in MERRA single-grid point data. In terms of vertical levels, MERRA multilevel averages show higher skill than single-level averages. Similar to ERA-int, levels located closer to the study site show considerably higher skill than more distant levels.

Figure 4.2 shows values of SS, at a monthly resolution, for increasing horizontal domains at the vertical levels with the highest skill of each reanalysis. Based on figure 4.2 the optimum horizontal domain is not as easily identified as based on the mean SS in figure 4.1, because optimum domain sizes differ for different months. In particular for some months, SS increases and decreases consecutively with increasing domain size.

The stepwise increase-decrease of SS with increasing domain sizes (cf. figure 4.2; due to monthly averaging of SS less but still evident in figure 4.1) can be explained by the geometry of the optimum domain analysis. In particular, the horizontal do-

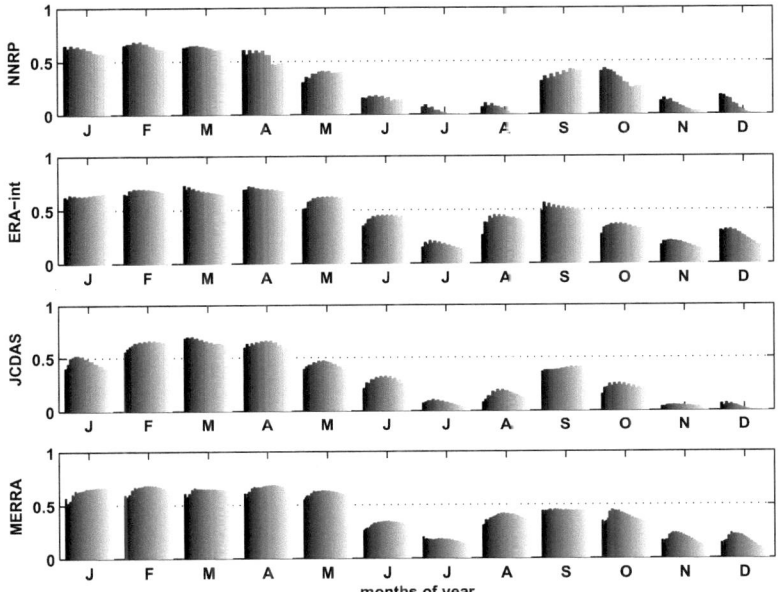

Figure 4.2: Values of SS (ordinate) for different months (abscissa) and for increasing horizontal domains (shadings of the bars) in the vertical level that shows the overall highest values of SS for NNRP (500 to 600 hPa average, top), ERA-int (550 hPa, second plot), JCDAS (500 to 600 hPa average, third plot), and MERRA (500 to 600 hPa average, bottom). Note that the domains are increasing from left to right, i.e., the first bar (darkest shading) refers to domain one, the second bar to domain two, etc. (for details about domain definitions the reader is refered to the text).

mains are increased by adding grid points to either western or eastern, and southern or northern sides of the domains, and in the following step, the domains are increased by adding grid point rows to the respective opposite sides. The stepwise increase-decrease of SS then results because grid points from one direction contain more information relevant to the local-scale data than grid points from the opposite direction. This indicates that horizontal domains arranged symmetrically around the study site are not necessarily the optimum domains, but shifting the domain towards synoptically more important regions can increase the skill. However, there is not one unique important flow direction for all months evident from figure 4.2, but the pattern of increase-decrease reverses several times throughout the year. More specifically,

Table 4.3: List of months (January 1, February 2, ...) for which increases (+) of SS for domain extensions towards southeast (SE) or northwest (NW) occur (for JCDAS southwest (SW) and northeast (NE) extensions, for MERRA south (S) or north (N) extensions). Increases for extensions towards SE on the same time imply decreases for extensions towards NW (and likewise for other directions).

	NNRP	ERA-int	JCDAS	MERRA
+SE (NNRP,ERA-int) +SW (JCDAS) +S (MERRA)	5, 7, 8, 9, 10, 11	7, 8, 9	5, 7, 8, 9	-
+NW (NNRP,ERA-int) +NE (JCDAS) +N (MERRA)	1, 4, 6	1, 3, 12	1, 2, 3, 4, 10	1, 3, 4, 6, 10

table 4.3 gives a summary of months where increases (decreases) for north or south (combined with west or east) extensions occur (note that for MERRA, the analysis is shown only for north or south extensions, because, due to the MERRA grid point geometry, West and East extensions occur simultaneously).

The different reanalyses show similar patterns of increase-decrease for similar months (cf. table 4.3). E.g. with northward extensions, for all reanalyses values of SS increase in January, for most reanalyses in March and April, for some in October. With southward extensions of the averaging domain, values of SS never increase for MERRA, but in July, August, September for all the other reanalyses. This is in some respects in accordance with the findings of *Georges* (2005), who performs a seasonal analysis of the tropospheric flow in several levels in the Cordillera Blanca. Even if *Georges* (2005) identifies northeasterly flow prevailing all year round, he finds that during humid conditions (especially January to March) the flow is more northerly than during dry conditions (especially June to August).

Table 4.4 shows a summary of the optimum scale analysis for all reanalyses. NNRP, most notably, show the smallest amount of grid points to be averaged for the optimum domain of all reanalyses (4 grid points in two levels, thus 8 grid points in total), and therefore the optimum scale is comparably small, even if the minimum scale of NNRP is the largest of all reanalyses. ERA-int is the only reanalysis where single-level data show the highest skill (i.e., the level located closest to the study site) and the optimum domain of ERA-int comprises relatively few grid points (16 in total). The resulting optimum scale for ERA-int is in fact the smallest of all reanalyses with

Table 4.4: Results of skill assessment for all reanalyses considered in this study: skill scores (annual mean and monthly maximum) obtained for the optimum domain, amount of grid points included in the optimum domain (in brackets: amount of grid points in the latitudinal×longitudinal×vertical direction), scale of the optimum domain (horizontally, in degrees), and optimum pressure level (pl).

	NNRP	ERA-int	JCDAS	MERRA
SS	0.36 (Feb:0.66)	0.48 (Apr:0.71)	0.35 (Mar:0.66)	0.46 (Feb:0.67)
Grid points	8 (2×2×2)	16 (4×4×1)	49 (7×7×2)	462 (14×11×3)
Scale	5×5	2.88×2.88	8.75×8.75	7×7.3
Pl	500 600	550	500,600	500,550,600

2.88°. In the case of JCDAS, more grid points to be averaged are required for the optimum domain, resulting the largest optimum scale of all reanalyses with 8.75°. As mentioned earlier for MERRA, of all reanalyses, the largest amount of grid points to be averaged compose the optimum domain. As MERRA have the smallest minimum scale, the optimum scale for MERRA is still smaller than for JCDAS, but larger than for NNRP and ERA-int.

To sum up, for all reanalyses the optimum domain is achieved with data from the pressure levels located close to the study site. However in terms of optimum horizontal scale, or optimum amount of grid points to be averaged, respectively, results widely vary for the different reanalyses from 2.88° (ERA-int) to 8.75° (JCDAS). Results in table 4.4 are further discussed in the next section.

4.4.2 Comparing the skill of different reanalyses

Table 4.4 and figure 4.3 summarize the performances of the four reanalyses with the optimum domain used as predictor. Of all reanalyses, ERA-int shows the highest mean value of SS. I.e., averaged over all calendar months, ERA-int reproduce almost half (48%) of daily air temperature variations in the Cordillera Blanca. Whereas MERRA shows comparably high average skill like ERA-int (mean $SS = 0.46$), both NNRP and JCDAS show considerably lower skill, reflecting only about a third of the local-scale air temperature variance in the Cordillera Blanca ($SS = 0.36$ for NNRP and $SS = 0.35$ for JCDAS).

For the definition of seasons in the Cordillera Blanca used hereafter please refer to *Niedertscheider* (1990) and *Juen* (2006). The overall highest values of SS result in the wet season, i.e.: in February (NNRP and MERRA), April (ERA-int), and

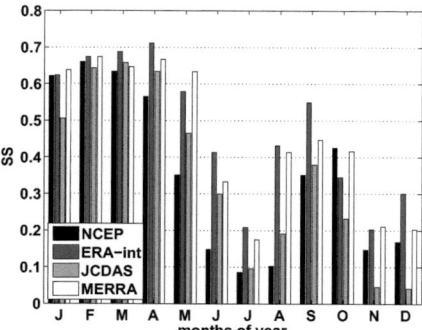

Figure 4.3: Values of SS for each calendar month (from left to right) and the different reanalyses (shadings of the bars).

March (JCDAS). In April, ERA-int reflects 71% of local-scale, daily air temperature variations in the Cordillera Blanca. A second maximum of SS occurs in the intermediate season September to October. The lowest values of SS are in the core dry season (especially July), but also in the intermediate season November to December. The seasonality of SS evident in figures 4.2 and 4.3 has already been discussed by *Hofer et al.* (2011) for NNRP. High values of SS in the wet season are attributed to strong, spatially homogenous synoptic forcing in the Cordillera Blanca during these months (e.g., related to the Madden-Julian Oscillation, *Madden and Julian*, 1994). By contrast, low values of SS during the dry season are due to weak synoptic activity combined with radiation-dominated thermal fluctuations triggered by local-scale orographic effects (note that during the dry season, atmospheric variability is generally smaller than during the rest of the year, *Hofer et al.*, 2011). Beyond these synoptic considerations, small values of SS potentially also derive from a larger probability of radiation errors affecting air temperature measurements during the dry season, than during the wet season.

ERA-int show the highest values of SS during the wet season, but also during the dry season SS never drops below 0.2. Whereas JCDAS show comparably high skill like the other reanalyses during the wet season, the performance during the dry season is considerably lower. In particular during the intermediate months November and December, values of SS of JCDAS are very low ($SS < 0.05$). As mentioned earlier MERRA show comparably high skill as ERA-int, but the second generation reanalysis JCDAS show almost lower skill than the first generation reanalysis NNRP.

Since NNRP and JCDAS both include 3D-Var, whereas ERA-int and MERRA include 4D-Var, this indicates that important improvements in the reanalysis systems derive to a lesser degree from lessons learned from earlier reanalysis projects (e.g, a better treatment of observational input), but rather from the substantial progress in the data assimilation systems and computing power over time (i.e., the transfer from 3D-Var to 4D-Var).

4.5 Conclusions

4.5.1 Results confined to the case study

We have not assessed whether skill and optimum scales of the reanalyses found in this study are transferable to regions outside the Cordillera Blanca, or to different variables. Here, we summarize important results confined to the assessed case study.

In terms of air temperature predictors in the Cordillera Blanca, ERA-int show the highest skill of all considered reanalses. Whereas MERRA show comparably high skill, JCDAS and NNRP show considerably lower skill. More specifically, even if all reanalyses perform relatively well for wet-season months, differences in skill between the different reanalyses are evident especially during the dry-season months, and the intermediate-season months November and December.

Regarding the optimum scale analysis, of all reanalyses ERA-int show the smallest optimum scale, with $2.88°$ (remember that this is 4 times the ERA-int grid distance, or minimum scale). In the case of NNRP, most notably, the optimum scale is only twice the minimum scale. This implies the fewest amount of grid points to be averaged for NNRP of all reanalyses. By contrast for MERRA, the ratio between optimum scale and minimum scale is 14 and is thus the largest of all considered reanalyses. This is an interesting result given that NNRP have the largest, and MERRA the smallest minimum scale of the reanalyses.

In terms of vertical levels, all reanalyses show the highest skill when data from pressure levels close to the study site are used, and vertical averaging hardly yields better results. For all reanalyses, the skill drops considerably if data are used from pressure levels more than $100hPa$ away from the study site.

4.5.2 General recommendations

Here we summarize conclusions to be generalized beyond the assessed case study. Even if results of the optimum scale analysis largely vary for different reanalyses, we find overall that the increase of skill with increasing domain is more important than the decrease of skill after reaching the maximum. We thus conclude that there are less negative effects when averaging over too large domains, than when too small domains are considered. Therefore, to gain the most reliable information from reanalyses, we can generally recommend horizontal grid point averaging rather than using single grid points. Vertical averaging, by contrast, shows no significant increase in skill. Including data from pressure levels located distant from the study site even lowers the skill considerably, for all reanalyses.

The minimum scale is not necessarily a good indicator for the optimum scale of large-scale model output (e.g., example of NNRP and MERRA). Regarding the skill assessment, we find that reanalyses that include 4D-Var data assimilation systems show notably higher performance than reanalyses with 3D-Var systems.

Finally, we like to point out that the analysis performed in this study can easily be repeated in different regions, or for other target variables, as long as a few-years observational data set is available. Because of the cross-validation procedure, the skill assessment is especially suited for short-term, high-resolution time series, with focus on inter-annual and intra-seasonal (day-to-day) variability.

BIBLIOGRAPHY

BIBLIOGRAPHY

Akaike, H. (1973), Information theory and an extension of the maximum likelihood principle, in *Second International Symposium on Information Theory*, pp. 267–281, Akademiai Kiado.

Ames, A. (1998), A documentation of glacier tongue variations and lake developement in the Cordillera Blanca, Peru, *Zeitung für Gletscherkunde und Glazialgeologie*, *34*(1), 1–36.

Baraer, M., J. M. McKenzie, B. G. Mark, J. Bury, and S. Knox (2009), Characterizing contributions of glacier melt and groundwater during the dry season in a poorly gauged catchment of the Cordillera Blanca (Peru), *Advances in Geosciences*, *22*, 41–49, doi:10.5194/adgeo-22-41-2009.

Benestad, R. (2001), A comparison between two empirical downscaling strategies, *International Journal of Climatology*, *21*(13), 1645–1668.

Benestad, R. E., I. Hanssen-Bauer, and C. Deliang (2008), *Empirical-statistical downscaling*, World Scientific, Singapore.

Bengtsson, L., and J. Shukla (1988), Integration of space and in situ observations to study global climate change, *Bulletin of the American Meteorological Society*, *69*(10), 1130–1143.

Bengtsson, L., S. Hagemann, and K. I. Hodges (2004), Can climate trends be calculated from reanalysis data?, *Journal of Geophysical Research*, *109*(D11111), doi: 0.1029/2004JD004536.

Bey, I., D. J. Jacob, and R. M. e. a. Yantosca (2001), Global modeling of tropospheric chemistry with assimilated meteorlogy: model description and evaluation, *Journal of Geophysical Research*, *106*, 23,073–23,095.

Bosilovich, M. G. (2008), Nasa's Modern Era Retrospective-analysis for research and applications: Integrating earth observations.

Bosilovich, M. G., J. Chen, F. R. Robertson, and R. F. Adler (2008), Evaluation of global precipitation in reanalyses, *Journal of Applied Meteorology and Climatology*, *47*(9), 2279–2299, doi:10.1175/2008JAMC1921.1.

Brinkmann, W. A. R. (2002), Local versus remote grid points in climate downscaling, *Climate Research*, *21*(1), 27–42.

Bury, J., B. Mark, J. McKenzie, A. French, M. Baraer, K. Huh, M. Zapata Luyo, and R. Gomez Lopez (2011), Glacier recession and human vulnerability in the Yanamarey watershed of the Cordillera Blanca, Peru, *Climatic Change*, *105*, 179–206, 10.1007/s10584-010-9870-1.

Carey, M. (2005), Living and dying with glaciers: people's historical vulnerability to avalanches and outburst floods in Peru, *Global and Planetary Change*, *47*, 122–134.

Carey, M. (2010), *In the Shadow of Melting Glaciers. Climate Change and Andean Society*, 288 pp., Oxford University Press.

Cavazos, T., and B. C. Hewitson (2005), Performance of NCEP-NCAR reanalysis variables in statistical downscaling of daily precipitation, *Climate Research*, *28*, 95–107.

Chevallier, P., B. Pouyaud, W. Suarez, and T. Condom (2011), Climate change threats to environment in the tropical Andes: glaciers and water resources, *Regional Environmental Change*, *11*, 179–187, 10.1007/s10113-010-0177-6.

Christensen, J., et al. (2007), *Regional Climate Projections*, Cambridge University Press, Cambridge, United Kingdom and New York, NY, USA.

Christensen, J. H., T. R. Carter, and F. Giorgi (2002), PRUDENCE employs new methods to assess European climate change, *EOS Transactions*, *83*, 147–147, doi: 10.1029/2002EO000094.

Davies, H. C. (1976), A lateral boundary formulation for multi-level prediction models, *Quarterly Journal of the Royal Meteorological Society*, *102*(432), 405–418, doi: 10.1002/qj.49710243210.

Dessler, A. E., and S. M. Davis (2010), Trends in tropospheric humidity from reanalysis systems, *Journal of Geophysical Research*, *115*(D19127), doi: 10.1029/2010JD014192.

DOE, D. o. t. E. (1996) Review of the potential effects of climate change in the united kingdom, *Tech. rep.*, HMSO, London.

Favier, V., P. Wagnon, J.-P. Chazarin, L. Maisincho, and A. Coudrain (2004), One-year measurements of surface heat budget on the ablation zone of Antizana Glacier 15, Equadorian Andes. *Journal of Geophysical Research*, *109*(D18105).

Fowler, H., S. Blenkisop, and C. Tebaldi (2007), Review: Linking climate change modelling to impacts studies: recent advances in downscaling techniques for hydrological modelling, *International Journal of Climatology*, *27*, 1547–1578.

Früh, B., J. W. Schipper, A. Pfeiffer, and V. Wirth (2006), A pragmatic approach for downscaling precipitation in alpine-scale complex terrain, *Meteorologische Zeitschrift*, *15*(16), 631–646.

Garreaud, R., M. Vuille, and A. Clement (2003), The climate of the Altiplano: observed current conditions and mechanisms of past changes, *Palaeogeography, Palaeoclimatology, Palaeoecology*, *194*(5-22).

Garreaud, R., M. Vuille, R. Compagnucci, and J. Marengo (2009), Present-day South American climate, *Palaeogeography, Palaeoclimatology, Palaeoecology*, *281*, 180–195.

Georges, C. (2002), Ventilated and unventilated air temperature measurements for glacier-climate studies on a tropical high mountain site, *Journal of Geophysical Research*, *107*(24).

Georges, C. (2004), 20th-century glacier fluctuations in the tropical Cordillera Blanca, Peru, *Arctic, Antarctic, and Alpine Research*, *36*(1), 100–107.

Georges, C. (2005), Recent glacier fluctuations in the tropical Cordillera Blanca and aspects of the climate forcing, Ph.D. thesis, Leopold-Franzens University Innsbruck.

Glahn, H. R., and D. A. Lowry (1972), The use of model output statistics (MOS) in objective weather forecasting, *Journal of Applied Meteorology*, *11*, 1203–1211.

Grotch, S. L., and M. C. MacCracken (1991), The use of global climate models to predict regional climatic change, *Journal of Climate*, *4*, 286–303.

Hannachi, A., I. T. Jolliffe, and D. B. Stephenson (2008), Empirical orthogonal functions and related techniques in atmospheric science: A review, *International Journal of Climatology*, *27*, 1119–1152.

Hastenrath, S. (1978), Heat budget measurements on Quelccaya Ice Cap, Peruvian Andes, *Journal of Glaciology*, *20*(82), 85–97.

Hertig, E., and J. Jacobeit (2008), Assessments of mediterranean precipitation changes for the 21st century using statistical downscaling techniques, *International Journal of Climatology*, *28*, 1025–1045.

Hewitson, B., and R. Crane (2006), Consensus between GCM climate change projections with empirical downscaling: precipitation downscaling over South Africa, *International Journal of Climatology*, *26*, 1315–1337.

Hill, G. E. (1968), Grid telescoping in numerical weather prediction, *Journal of Applied Meteorology*, *7*, 29–38, doi:10.1175/1520-0450.

Hofer, M. (2007), Statistical downscaling of NCEP/NCAR reanalysis data to air temperature and specific humidity above an outer tropical glacier surface; Artesonraju (Peru), Master's thesis, Leopold Franzens University, Innsbruck.

Hofer, M., T. Mölg, B. Marzeion, and G. Kaser (2010), Empirical-statistical downscaling of reanalysis data to high-resolution air temperature and specific humidity above a glacier surface (Cordillera Blanca, Peru), *Journal of Geophysical Research*, *115*(D12120), 15.

Hofer, M., B. Marzeion, and T. Mölg (2011), Skill assessment of NCEP/NCAR reanalysis data for daily air temperature on a glaciated mountain range (Peru), *Journal of Climate, submitted*.

Hughes, J. P., and P. Guttorp (1994), A class of stochastic models for relating synoptic atmospheric patterns to regional hydrologic phenomena, *Water Resources Research*, *30*(5), 1535–1546, doi:10.1029/93WR02983.

Huth, R. (2004), Sensitivity of local daily temperature change estimates to the selection of downscaling models and predictors, *Journal of Climate, 17*.

Jones, P. D., and A. Moberg (2003), Hemispheric and large-scale surface air temperature variations: An extensive revision and an update to 2001, *Journal of Climate*, *16*(2), 206–223, doi:10.1175/1520-0442.

Juen, I. (2006), Glacier mass balance and runoff in the Cordillera Blanca, Peru, Ph.D. thesis.

Juen, I., C. Georges, and G. Kaser (2007), Modelling observed and future runoff from a glacierized tropical catchment (Cordillera Blanca Peru), *Global and Planetary Change, 59*, 37–48.

Juen, I., P. Wagnon, G. Kaser, and J. Gomez (2011) Seasonal variation of energy balance fluxes on Glaciar Artesonraju in the tropical Cordillera Blanca, Peru, *Geophysical Research Letters, submitted*.

Kalnay, E., et al. (1996), The NCEP/NCAR 40-year reanalysis project, *Bulletin of the American Meteorological Society, 77*(3), 437–471.

Karl, T., S. Hassol, S. Miller, and W. Murray (2006), Temperature trends in the lower atmosphere: Steps for understanding and reconciling differences.. *Tech. rep.*

Kaser, G. (1995), Some notes on the behaviour of tropical glaciers., *Bulletin de l'Institut Francais d'Etudes Andines, 24*(3), 671 – 681.

Kaser, G., and C. Georges (1997), Changes in the equilibrium line altitude in the tropical Cordillera Blanca (Peru) between 1930 and 1950 and their spatial variations., *Annals of Glaciology, 24*, 344–349.

Kaser, G., and H. Osmaston (2002), *Tropical Glaciers*, International Hydrology Series, Cambridge University Press, Cambridge, UK.

Kaser, G., A. Ames, and M. Zamora (1990), Glacier fluctuations and climate in the Cordillera Blanca, Peru, *Annals of Glaciology, 14*, 136–140.

Kaser, G., S. Hastenrath, and A. Ames (1996), Mass balance profiles on tropical glaciers, *Zeitung für Gletscherkunde und Glazialgeologie, 32*, 75–81.

Kaser, G., I. Juen, C. Georges, J. Gomez, and W. Tamayo (2003), The impact of glaciers on the runoff and the reconstruction of mass balance history from hydrological data in the tropical Cordillera Blanca, Peru, *Journal of Hydrology*, *282*(1-4), 130–144.

Kaser, G., C. Georges, I. Juen, and T. Mölg (2005), *Low-latitude glaciers: Unique global climate indicators and essential contributors to regional fresh water supply. A conceptual approach.*, vol. 23, pp. 185–196., Kluwer, New York.

Kaser, G., M. Grosshauser, and B. Marzeion (2010), The contribution potential of glaciers to water availability in different climate regimes, *Proceedings of the National Academy of Sciences*, *107*, 20,223–20,227.

Kazutoshi, O., et al. (2007), The JRA-25 reanalysis, *Journal of the Meteorological Society of Japan*, *85*(3), 369–432.

Kim, J.-W., J.-T. Chang, N. Baker, D. Wilks, and W. Gates (1984), The statistical problem of climate inversion: Determination of the relationship between local and large-scale climate, *Monthly Weather Review*, *112*(10), 2069–2077.

Klein, W. H., and H. R. Glahn (1974), Forecasting local weather by means of model output statistics, *Bulletin American Meteorological Society*, *55*(10), 1217–1227.

Klein, W. H., B. M. Lewis, and E. I. (1959), Objective prediction of five-day mean temperature during winter, *Journal of Meteorology*, *16*(6), 672–682.

Kuhn, M. (1989), *The response of the equilibrium line altitude to climate*, pp. 407–417, Glaciology and quaternary geology, Kluwer Academic Publishers, Dordrecht, The Netherlands.

Kunsch, H. R. (1989), The jackknife and the bootstrap for general stationary observations, *The Annals of Statistics*, *17*(3), 1217–1241.

Kutzbach, J. E. (1967), Empirical eigenvectors of sea-level pressure, surface temperature and precipitation complexes over north america, *Journal of Applied Meteorology*, *6*, 791–802.

Lorenz, E. N. (1969), Atmospheric predictability as revealed by naturally occurring analogues, *Journal of the Atmospheric Sciences*, *26*, 636–646.

Madden, R. A. (1976), Estimates of the natural variability of time-averaged sea-level pressure, *Monthly Weather Review*, *104*(7), 942–952, doi:10.1175/1520-0493(1976)104.

Madden, R. A. (1979), A simple approximation for the variance of meteorological time averages, *Journal of Applied Meteorology*, *18*.

Madden, R. A., and P. R. Julian (1994), Observations of the 40 to 50-day tropical oscillation: A review, *Monthly Weather Review*, *122*(5), 814–837, doi:10.1175/1520-0493(1994)122.

Maraun, D., et al. (2010), Precipitation downscaling under climate change. recent developments to bridge the gap between dynamical models and the end user, *Reviews of Geophysics*, *48*.

Mark, B. G., and G. O. Seltzer (2003), Tropical glacier meltwater contribution to stream discharge: a case study in the Cordillera Blanca, Peru, *Journal of glaciology*, *49*(165), 271–281.

Mark, B. G., J. Bury, J. M. McKenzie, A. French, and M. Baraer (2010), Climate change and tropical Andean glacier recession: Evaluating hydrologic changes and livelihood vulnerability in the Cordillera Blanca, Peru, *Annals of the Association of American Geographers*, *100*(4), 794805.

Martin, E., B. Timbal, and E. Brun (1996), Downscaling of general circulation model outputs: simulation of the snow climatology of the french alps and sensitivity to climate change, *Climate Dynamics*, *13*, 45–56, 10.1007/s003820050152.

Matulla, C. (2005), Regional, seasonal and predictor-optimized downscaling to provide groups of local scale scenarios in the complex structured terrain of Austria, *Meteorologische Zeitschrift*, *14*(1), 31–45(15).

McFarlane, N. (2011), Parameterizations: representing key processes in climate models without resolving them, *Wiley Interdisciplinary Reviews: Climate Change*, *2*(4), 482–497, doi:10.1002/wcc.122.

Michaelsen, J. (1987), Cross-validation in statistical climate forecast models, *Journal of Climate and Applied Meteorology*, *26*, 1589–1600.

Miguez-Macho, G., G. L. Stenchikov, and A. Robock (2004), Spectral nudging to eliminate the effects of domain position and geometry in regional climate model simulations, *Journal of Geophysical Research*, *109*(D13104) doi:10.1029/2003JD004495.

Mölg, T., and D. R. Hardy (2004), Ablation and associated energy balance of a horizontal glacier surface on Kilimanjaro, *Journal of Geophysical Research (Atmospheres)*, *109*, D16,104, doi:10.1029/2003JD004338.

Mölg, T., and G. Kaser (2011), A new approach to resolving climate-cryosphere relations: Downscaling climate dynamics to glacier-scale mass and energy balance without statistical scale linking, *Journal of Geophysical Research*, accepted.

Mölg, T., N. Cullen, and G. Kaser (2009), Solar radiation, cloudiness and longwave radiation over low-latitude glaciers: Implications for mass balance modeling, *Journal of Glaciology*, *55*, 292–302.

Mölg, T., M. Grohauser, A. Hemp, M. Hofer, and B. Marzeion (2011), Is there additional forcing of mountain glacier loss through land cover change?, *Nature Geoscience, submitted*.

Murphy, A. H. (1988), Skill scores based on the mean square error and their relationships to the correlation coefficient, *Monthly Weather Review*, *116*(12), 2417–2424, doi:10.1175/1520-0493(1988)116.

Niedertscheider, J. (1990), Untersuchungen zur Hydrographie der Cordillera Blanca (Peru), Master's thesis, Leopold Franzens University, Innsbruck.

Paterson, W. S. P. (1994), *The physics of glaciers*, 3 ed., Butterworth-Heinemann, Oxford.

Prömmel, K., B. Geyer, J. M. Jones, and M. Widmann (2010), Evaluation of the skill and added value of a reanalysis-driven regional simulation for alpine temperature, *International Journal of Climatology*, *30*(5), 760–773, doi:10.1002/joc.1916.

Qian, B., H. Hayhoe, and S. Gameda (2005), Evaluation of the stochastic weather generators LARS-WG and AAFC-WG for climate change impact studies, *Climate Research*, *29*, 3–21.

Racoviteanu, A. E., Y. Arnaud, M. W. Williams, and J. Ordoñez (2008), Decadal changes in glacier parameters in the Cordillera Blanca, Peru, derived from remote sensing, *Journal of Glaciology*, *54*(186), 499–510.

Radic, V., and R. Hock (2006), Modeling future glacier mass balance and volume changes using era-40 reanalysis and climate models: A sensitivity study at storglaciren, sweden, *Journal of Geophysical Research*, *111*.

Räisänen, J., and J. S. Ylhäisi (2011), How much should climate model output be smoothed in space?, *Journal of Climate*, *24*(3), 867–880, doi: 10.1175/2010JCLI3872.1.

Reichert, B. K., and L. Bengtsson (2002), Recent glacier retreat exceeds internal variability, *Journal of climate*, *15*(23), 3069–3081.

Richardson, C. W. (1981), Stochastic simulation of daily precipitation, temperature, and solar radiation, *Water Resources Research*, *17*(1), 182190, doi: 10.1029/WR017i001p00182.

Rood, R. B., and M. G. Bosilovich (2009), *Reanalysis: data assimilation for scientific investigation of climate*, Springer.

Salzmann, N. (2006), The use of results from regional climate models for local-scale permafrost modelling in complex high-mountain topography - possibilities, limitations and challenges for the future, Ph.D. thesis.

Schoof, J., and S. Pryor (2001), Downscaling temperature and precipitation: a comparison of regression-based methods and artificial neural networks, *International Journal of Climatology*, *21*, 773–790.

Schubert, S., and A. Henderson-Sellers (1997), A statistical model to downscale local daily temperature extremes from synoptic-scale atmospheric circulation patterns in the Australian region, *Climate Dynamics*, *13*, 223–234.

Silverio, W., and J.-M. Jaquet (2005), Glacial cover mapping (1987-1996) of the Cordillera Blanca (Peru) using satellite imagery, *Remote Sensing of the Environment*, *95*, 342–350.

Simmons, A., S. Uppala, D. Dee, and S. Kobayashi (2007), ERA-Interim: new ECMWF reanalysis products from 1989 onwards.

Simmons, A. J., and P. D. Jones (2004), Comparison of trends and low-frequency variability in CRU, ERA-40, and NCEP/NCAR analyses of surface air temperature, *Journal of Geophysical Research*, *109*(D24115), doi:10.1029/2004JD005306.

Trenberth, K., et al. (2007), Observations: Surface and atmospheric climate change, Tech. rep.

Trenberth, K. E., and J. G. Olson (1988), An evaluation and intercomparison of global analyses from the National Meteorological Center and the European Centre for Medium Range Weather Forecasts, *Bulletin of the American Meteorological Society*, *69*(9), 1047–1057, doi:10.1175/1520-0477(1988)069.

Trenberth, K. E., D. P. Stepaniak, J. W. Hurrell, and M. Fiorino (2001), Quality of reanalyses in the tropics, *Journal of Climate*, *14*(7), 11.

Uppala, S., et al. (2005), The ERA-40 re-analysis, *Quarternary Journal of the Royal Meteorological Society*, *131*, 2961–3012.

Von Storch, H. (1999), On the use of inflation in statistical downscaling, *Journal of Climate*, *12*, 3505–3506.

Von Storch, H., and F. Zwiers (2001), *Statistical analysis in climate research*, 484 pp., Cambridge University Press, Cambridge, UK.

Von Storch, H., E. Zorita, and U. Cubasch (1993), Downscaling of global climate change estimates to regional scales: An application to iberian rainfall in wintertime, *Journal of Climate*, *6*(6), 1161–1171, doi:10.1175/1520-0442(1993)006.

Von Storch, H., B. Hewitson, and L. Mearns (2000a), Review of empirical downscaling techniques, in *Regional climate development under global warming*, edited by T. Iversen and B. A. K. Hoiskar, General Technical Report 4.

Von Storch, H., H. Langenberg, and F. Feser (2000b), A spectral nudging technique for dynamical downscaling purposes, *Monthly Weather Review*, *128*, 3664–3673.

Vuille, M., and F. Keimig (2004), Interannual variability of summertime convective cloudiness and precipitation in the central Andes derived from ISCCP-B3 data, *Journal of Climate, 17*, 3334–3348.

Vuille, M., G. Kaser, and I. Juen (2008), Glacier mass balance variability in the Cordillera Blanca, Peru and its relationship to climate and large scale circulation, *Global and planetary change, 62*, 14–28.

Wagnon, P., P. Ribstein, B. Francou, and B. Pouyaud (1999), Annual cycle of energy balance of Zongo glacier, Cordillera Real, Bolivia, *Journal of Geophysical Research, 104*(D4), 3907–3923.

Warner, T. T., R. A. Peterson, and R. E. Treadon (1997), A tutorial on lateral boundary conditions as a basic and potentially serious limitation to regional numerical weather prediction, *Bulletin of the American Meteorological Society, 78*(11), 2599–2617.

Widmann, M. (2005), One-dimensional CCA and SVD, and their relationship to regression maps, *Journal of Climate, 18*(14), 2785–2792.

Widmann, M., C. S. Bretherton, and E. Salath (2003), Statistical precipitation downscaling over the northwestern United States using numerically simulated precipitation as a predictor, *Journal of Climate, 16*(5), 799–816.

Wilby, R., C. Dawson, and E. Barrow (2002), SDSM - a decision support tool for the assessment of regional climate change impacts, *Environmental and Modelling Software, 17*, 145–157.

Wilks, D. S. (1997), Resampling hypothesis tests for autocorrelated fields, *Journal of Climate, 10*(1), 65–82.

Wilks, D. S. (2006), *Statistical methods in the atmospheric sciences, International Geophysics Series*, vol. 91, 2 ed., Academic Press.

Wilks, D. S., and R. L. Wilby (1999), The weather generation game: a review of stochastic weather models, *Progress in Physical Geography, 23*(3), 329–357.

Willamson, D. L., and R. Laprise (2000), *Numerical modeling of the global atmosphere in the climate system*, chap. Numerical approximations for global atmospheric GCMs, pp. 147–219, Kluwer Academic.

Xoplaki, E., J. Gonzalez-Rouco, J. Luterbacher, and H. Wanner (2003), Mediterranean summer air temperature variability and its connection to the large-scale atmospheric circulation and ssts, *Climate dynamics, 20*(7-8), 723–739.

Zorita, E., and H. Von Storch (1997), A survey of statistical downscaling techniques, *Tech. rep.*, GKKS.

Zorita, E., J. P. Hughes, D. P. Lettemaier, and H. von Storch (1995), Stochastic Characterization of Regional Circulation Patterns for Climate Model Diagnosis and Estimation of Local Precipitation., *Journal of Climate*, 8, 1023–1042, doi: 10.1175/1520-0442.

LIST OF FIGURES

1.1 Map of the Cordillera Blanca with main glaciers and the Rio Santa water shed (the indicated positions of the glacier Artesonraju and the lake Querococha are referred to in chapter II). 9

2.1 The role of downscaling techniques to transfer atmospheric data with high availability in time and space (abscissa), but low spatial resolution (ordinate), to the higher spatial resolutions required in climate research. In this figure we use the more general term "global climate model output", but it is true also for reanalysis data. 23

2.2 NCEP model topography over South America. The crosses are grid points. The black line connects the highest locations in the NCEP model topography. The black rectangle shows the horizontal area finally included in the ESD model. 26

2.3 Adjusted correlation of determination (r^2) between the leading 10 PCs of the fields a, s, h, u, and v, for the predictands air temperature and specific humidity. The four bars of each variable correspond to the vertical locations of the predictor fields: 1000, 800, 600 and 400hPa levels from left to right. 32

2.4 Illustration of the PC screening procedure applied in this study using the example of the monthly/hourly case December/13LT, for the predictands air temperature (left) and specific humidity (right) and the single-field predictor PCs of a_{1000} and s_{1000}, respectively. Top panel: mse_f (solid lines) and mse_h (dashed lines) as a function of k. Lower panel: $AIC(k)$ (arbitrary units). 33

2.5 Skill scores of the month/hour-transfer functions estimated by double cross-validation for the predictand air temperature (left) and specific humidity (right), single-field (crosses) and mixed-field (circles) predictors. 35

2.6 Diurnal cycles of air temperature (top) and specific humidity (bottom) at AWS: observations (dashed), ESD model hindcast (thin), and reference model (solid); a case study in April 2004 (1LT in the first x-tick means 1 Peruvian Local Time). 37

2.7 Daily mean air temperature (top) and specific humidity (bottom) at AWS: observations (dashed), ESD model hindcast (thin), and reference model (solid) over the calibration period (March 2004 to May 2006). Periods with missing data are grey shaded. 39

2.8 Monthly means of air temperature (top) and specific humidity (bottom): observations (dashed), ESD model hindcast (thin), and reanalysis grid point (solid) in the calibration period (March 2004 to May 2006). Periods with missing data are grey shaded. 40

2.9 Top: Annual means of ESD model forecast (dark), NCEP grid point values (thin) and observations at Querococha (solid grey, only air temperature) of air temperature (left) and specific humidity (right) for the period 1960 to 2008. El Niño (o) and La Niña (a) events are indicated in the abscissa. Bottom: ESD model forecast of annual mean air temperature (left) and specific humidity (right) for the period 1960 to 2008, with two different predictor fields applied for each variable: the single-field predictors a_{1000} for air temperature and s_{1000} for specific humidity (solid), and the mixed-field predictors including a_{1000}, s_{1000}, and u_{400} (black). The model training period (2004 to 2006) is grey shaded. 42

3.1 The map shows the Rio Santa watershed with the Cordillera Blanca mountain range and measurement sites (mentioned in the text). Also indicated is the 1990 glacier extend (grey shaded area) (*Georges*, 2004). 45

3.2 Schematic of the two classical statistical downscaling approaches MOS and PP, and options (*). The abbreviation T denotes training period, and F forecasting (or model application) period of the different approaches. The numberings of the different approaches 1-4 are referred to in the text. 47

3.3 Statistics of hourly air temperature time series at AWS1 (5000 m a.s.l.), AWS2 (4800 m a.s.l.), and AWS3 (4950 m a.s.l.), and 6-hourly values of the NCAR reanalysis data predictor $air500600$ (as defined in the text), for each month of the year (abscissa: January to December). Shown are the means (blue solid line) and the medians (red dashes). The edges of the thick bars are the 25th and the 75th percentiles. The thin bars extend to the most extreme data not considered as outliers, and the crosses are the outliers. The statistics are computed over the period of available measurements July 2006 to July 2010. 49

3.4 Example of skill estimation by cross-validation as described in section 3.3.3 for AWS1 predictands in January (top) and July (bottom). $y(t)$ (crosses-line) $\hat{y}(t)$ (solid grey), y_r (solid black), $y_V(cv)$ (star), $\hat{y}_V(cv)$ (red star), and $y_r(cv)$ (black star). 56

3.5 Medians (lines in within the boxes) of the downscaling model parameters α (top) and R_σ (bottom) estimated by cross-validation for the AWS1 predictands and all calendar months. The edges of the blue boxes are the 25th and the 75th percentiles. The black dashes (outside the boxes) extend to the most extreme data not considered as outliers. 57

3.6	Decorrelation time τ of daily air temperature observations at AWS1 (black), AWS2 (grey), and AWS3 (white) for each month of the year.	58
3.7	Values of SS for AWS1 (black), AWS2 (gray), and AWS3 (white) for the predictand $air500600$ estimated by cross-validation (as described in 3.3.3).	59
3.8	Averaged values of SS (black) and r^2 (white) over all months of the year for AWS1, AWS2, and AWS3 (from left to right).	60
3.9	SSs for different months (from left to right within each column, or black to white, respectively: January to December) and different time scales (from 1-day averages to 5-day averages). Note that, because here $n = 13$, versus $n > 100$ as in the previous sections, the values of SS for 1-daily means do not correspond to the results in figure 3.7. The black points connected by the bold line shows SS averaged over all months. The values are shown for the AWS1 data series. Negative values of SS are not shown.	61
3.10	Values of SS for each month (here for AWS1), and different predictors (from left to right): $airSFC$, slp, $hgt600$ (light blue), $air600$ (green), $air500$ (yellow), $air500600$, and $air18gp$.	63
3.11	r^2 for 6-hourly and daily (all-month) time series of AWS1 air temperature and the predictors (from left to right): $airSFC$, slp, $hgt600$, $air600$, $air500$, $air600500$, and $air18gp$.	64
4.1	Values of SS averaged over all calendar months for different vertical levels and combinations (different colors and lines) and increasing horizontal domains (from left to right), for NNRP (top left), ERA-int (top right), JCDAS (bottom left), and MERRA (bottom right) predictors. Please note that the scale on the abscissa changes for the different reanalyses, because of the different grid point spacings (the scale on the ordinate is kept fix).	75
4.2	Values of SS (ordinate) for different months (abscissa) and for increasing horizontal domains (shadings of the bars) in the vertical level that shows the overall highest values of SS for NNRP (500 to 600 hPa average, top), ERA-int (550 hPa, second plot), JCDAS (500 to 600 hPa average, third plot), and MERRA (500 to 600 hPa average, bottom). Note that the domains are increasing from left to right, i.e., the first bar (darkest shading) refers to domain one, the second bar to domain two, etc. (for details about domain definitions the reader is refered to the text).	77
4.3	Values of SS for each calendar month (from left to right) and the different reanalyses (shadings of the bars).	80

Figure

i want morebooks!

Buy your books fast and straightforward online - at one of world's fastest growing online book stores! Environmentally sound due to Print-on-Demand technologies.

Buy your books online at

www.get-morebooks.com

Kaufen Sie Ihre Bücher schnell und unkompliziert online – auf einer der am schnellsten wachsenden Buchhandelsplattformen weltweit! Dank Print-On-Demand umwelt- und ressourcenschonend produziert.

Bücher schneller online kaufen

www.morebooks.de

VDM Verlagsservicegesellschaft mbH
Heinrich-Böcking-Str. 6-8   Telefon: +49 681 3720 174   info@vdm-vsg.de
D - 66121 Saarbrücken   Telefax: +49 681 3720 1749 www.vdm-vsg.de

Printed by Books on Demand GmbH, Norderstedt / Germany